The Designer's AutoCAD 2000 Tutorial

Joan McLain-Kark

Virginia Polytechnic Institute and State University

2001

DAME
Thomson Learning

Australia • Canada • Denmark • Japan • Mexico • New Zealand • Philippines
Puerto Rico • Singapore • South Africa • Spain • United Kingdom • United States

The Designer's AutoCad 2000 Tutorial, by Joan McLain-Kark
Cover Design: Andrea P. Leggett
Cover Photos: © Corel Professional Photos. Images may have been combined and/or modified to produce final cover art.
Printer: Globus Printing

Printed in the United States of America
1 2 3 4 5 02 01 00 99

For more information contact Thomson Learning Custom Publishing, 5101 Madison Road, Cincinnati, Ohio, 45227, 1-800-355-9983 or find us on the Internet at http://www.custom.thomsonlearning.com

For permission to use material from this text or product, contact us by:
- **telephone: 1-800-730-2214**
- **fax: 1-800-730-2215**
- **web: http://www.thomsonrights.com**

Library of Congress Cataloging-in-Publication Data: 00-133327

ISBN 0-324-06411-X

This book is printed on acid-free paper.

Warning and Disclaimer

Trademarks

Table of Contents at a Glance

Part One: Basics

Part Two: Beginning AutoCAD

Part Three: Drafting

Part Four: 3-D Modeling

Table of Contents

Part Three: Drafting

Part Four: 3D Modeling

Preface

In the past, it was expected that every designer had manual drafting skills even if this task was assigned to others in a design firm. When CAD was introduced in the mid-1980s, most firms relied on a few CAD operators or designers, usually newly graduated, to draft construction drawings on their CAD systems. Gradually, designers expanded their CAD use to include space planning, scheduling, construction drawings, computer modeling, and rendering. Today, it is expected that every designer, from the top echelons of the company to the entry-level designer, should know CAD.

This book is intended for designers who want to learn AutoCAD, the most commonly used CAD software in interior design. It does not assume any prior knowledge of Computer-Aided Design (CAD) systems or knowledge of computer systems. It assumes only a fundamental knowledge of a microcomputer such as turning on the computer and using the keyboard and mouse.

This book should be useful to students in CAD courses intended for interior design students. The tutorials allow the students to learn AutoCAD quickly and provide meaningful examples that are relevant to design. By providing additional projects, the teacher can allow students to apply their newly acquired skills to portfolio projects.

This book is also useful as a reference for students after they take a CAD course when they use CAD in their regular studios. Procedures for layering and setting up a drawing can be looked up quickly and applied to their work.

Teachers of upper-level studios can use this tutorial to learn AutoCAD if they do not know it already. With this background, design studio instructors will better be able to develop projects where students use CAD. Just as all designers in a design firm need to know CAD, so should all teachers in an interior design program be expected to have this important design skill.

This book is also suitable for interior designers and facility planners needing to learn AutoCAD for their work or who may just want to refresh their CAD skills. It can be completed in approximately twenty hours, depending on the designer. The book may be especially useful to the top management and owners of design firms who supervise CAD designer, but who do not have expertise in AutoCAD themselves. This knowledge can enable them to understand the potential as well as limitations of this technology.

Acknowledgments

Many people have helped me with this book which has been developed over many years. First of all, I would like to thank the interior design students at Virginia Tech who have used the tutorial in its varied forms over the last 16 years. Students in my first year course, Design Drawing, during spring 2000 provided me with many ideas and suggestions on revising this book. I also would like to thank the faculty and administrators at Virginia Tech for their support and advice.

Lori Anthony, graduate teaching assistant, helped trouble-shoot the manuscript for technical and typographical errors. Her assistance in teaching CAD and helpful advice on this book are greatly appreciated.

I would like to thank both Lisa Waxman, Florida State University, and Stephanie Clemons, Colorado State University, for the idea to make a rug design to practice basic AutoCAD skills.

Dame Publications, Division of Thomson Learning, is appreciated for their commitment to help keep this tutorial current with the latest releases of AutoCAD. Art Francia, Dame founder and publisher, understands the needs of teachers and authors. I appreciate his flexibility and willingness to work with my schedule.

Finally, I am most thankful to my husband, Steve Kark, who has provided me with great inspiration from his own writing as a newspaper columnist and as an English faculty member at Virginia Tech. His love continues to sustain me.

About the Book

The book is composed of 16 tutorials organized into four parts. **Part One** presents an overview of AutoCAD and use of Windows 98 for managing drawing files. **Part Two** is a series of four tutorials which introduces the basic AutoCAD draw and edit commands by use of simple drawings. **Part Three** concerns how to space plan and draft two-dimensional drawings. First, students will use layering to make a building shell, furniture plan, lighting plan, and dimensioned drawing. Then paper space will be used to plot out different versions of these drawings. A drawing checklist and procedures is presented at this end of the section providing a helpful reference on projects. **Part Four** covers computer modeling with AutoCAD. Students will first make a room, add 3-D furniture, and then generate a perspective line drawing more commonly referred to as a wireframe drawing. Then, they will learn to render the computer model by adding lights, materials, and textures. Solid modeling is covered in the last tutorial which covers the greatly expanded solids editing features of AutoCAD 2000. Finally, a short section covers resources and tips for continued work with AutoCAD.

The book covers use of the menu bar, toolbars, and commands. Emphasis is placed on using the toolbar icons since this is the fastest way to enter commands. Command sequences are provided so the student can match these with what is displayed in AutoCAD. All instructions are listed in command sequences or lists printed in bold so students will see them easily. The beginning tutorials emphasize basic commands while the later exercises introduce more commands by applying AutoCAD to the types of drawings that designers and architects produce (furniture plans, elevations, and perspective drawings).

At the end of the first 6 tutorials are questions and practice exercises to help students reinforce their newly acquired skills. Lesser used commands and options are presented at the end of each tutorial for reference. As a consequence, the student will take about 20-25 hours to go through the tutorials and then be able to apply their skills to actual design projects.

Most of the tutorials have the following format:

- **Commands Learned** is a short of list of commands to be learned in the tutorial.

- **Overview** discusses what the students will learn and do in the tutorial.

- **Tutorial** is the largest proportion of each chapter and is the step-by-step tutorial which has the command sequences, lists of instructions, toolbar icons, and explanatory text. Almost every task is accompanied by an illustration of the way the drawing should appear before and after execution of the commands.

- **Short Answer/Discussion Questions** test students on what they learned.

- **Practice** suggests ways the student can practice their AutoCAD skills through projects or drawings. Practice sections are only included in the first 6 tutorials because the rest of the tutorials directly apply their AutoCAD skills to interior design examples.

- **Additional Work** covers little used options or commands which can be used for reference.

Teacher's Guide

The **Teacher's Guide for The Designer's AutoCAD 2000 Tutorial** is available to accompany this book. This guide contains several components:

- **Teaching/Learning Methods.** This section offers suggestions on demonstration techniques for each tutorial, covers possible problems students may encounter, and presents general organization of a CAD or CAD Studio course.

- **Sample Syllabi.** Sample syllabi are offered which cover different approaches to teaching a CAD course.

- **Project Outlines.** Sample project outlines show how the AutoCAD skills can be applied to several interior design projects for various levels of instruction.

About the Author

Joan McLain-Kark is Professor of Interior Design at Virginia Polytechnic Institute and State University, or more commonly known as Virginia Tech. She received her bachelors and masters degrees in housing and interior design from the University of Missouri-Columbia and her doctorate in human ecology from the University of Tennessee-Knoxville. McLain-Kark has taught AutoCAD to interior design students since 1984 making Virginia Tech one of the first interior design progams to offer AutoCAD instruction. She has taught interior design in five interior design programs. McLain-Kark practiced full-time in residential design and office planning.

McLain-Kark served as chair of the Computer Network of the Interior Design Educators Council from 1987-1990. She was a founding board member of the Design Communication Association. She is editor of the electronic **Journal of Design Communication** and former editor of the electronic **Journal of Computer-Aided Environmental Design and Education**. She is an editorial board member of the **Journal of Interior Design** and has edited three conference proceedings. In 1998, McLain-Kark was appointed to the Foundation for Interior Design Education Research's Research Council. In 1999, she was the keynote speaker on computing in design at Third Scientific Meeting of Interior Design in Kuwait and was interviewed on two television programs in Kuwait.

McLain-Kark has authored over 45 refereed articles and proceedings publications. Her previous books, **The Designer's AutoCAD Tutor,** was published by Van Nostrand Reinhold in 1991 and **The Designer's AutoCAD Release 14 Tutorial** by Dame Publications in 1998. She has won several awards and obtained over 30 grants to support CAD instruction and research at Virginia Tech. She taught CEU courses on AutoCAD for designers.

McLain-Kark's students have garnered 10 awards in national design competitions. Her first three doctoral students won the American Society of Interior Designers (ASID)/Polksy Academic Achievement Award for their research, the only prize awarded for theses or dissertations in interior design. Her graduate student advisees have also won numerous fellowships.

Joan McLain-Kark is Director of the Interior Design Futures Lab. The lab focuses on innovative use of computer technology including virtual reality in the design field. She is presently researching the use of virtual reality as a tool to improve the design process.

McLain-Kark lives with her husband, their two dogs, and cat on their farm on Guinea Mountain in the Appalachian mountains of southwestern Virginia.

Part One: Basics

Overview of AutoCAD

AutoCAD is the number-one selling microcomputer CAD software. This success can be attributed to the generic nature of the software, its large support of peripheral devices, and its open architecture. Open architecture means that a programmer can change the software. This has prompted the development of third-party software and products for AutoCAD which enables each field to have software to meet its needs.

Drawing Editor

The illustration (Fig. 1-1) shows the Drawing Editor as it appears when AutoCAD 2000 is newly installed. You use **commands** to tell AutoCAD what you want to do, such as drawing or erasing a line. The lines, circles, arcs, text, and other graphic images you draw are called **objects**. Once objects are drawn, they can be edited in a variety of ways such as moving, copying, or erasing. The drawing or graphic window is the area where you draw. At the top, there is a menu bar and immediately below is the Standard Toolbar containing common file, utility, and zoom commands. Below that is the Object Properties Toolbar containing layer, color, linetype, and lineweight information. Toolbars can be placed just about anywhere by the user. At the bottom of the screen is the command prompt area where you enter commands. Below that is the status line where you can check on the coordinate location, snap, grid, or other settings.

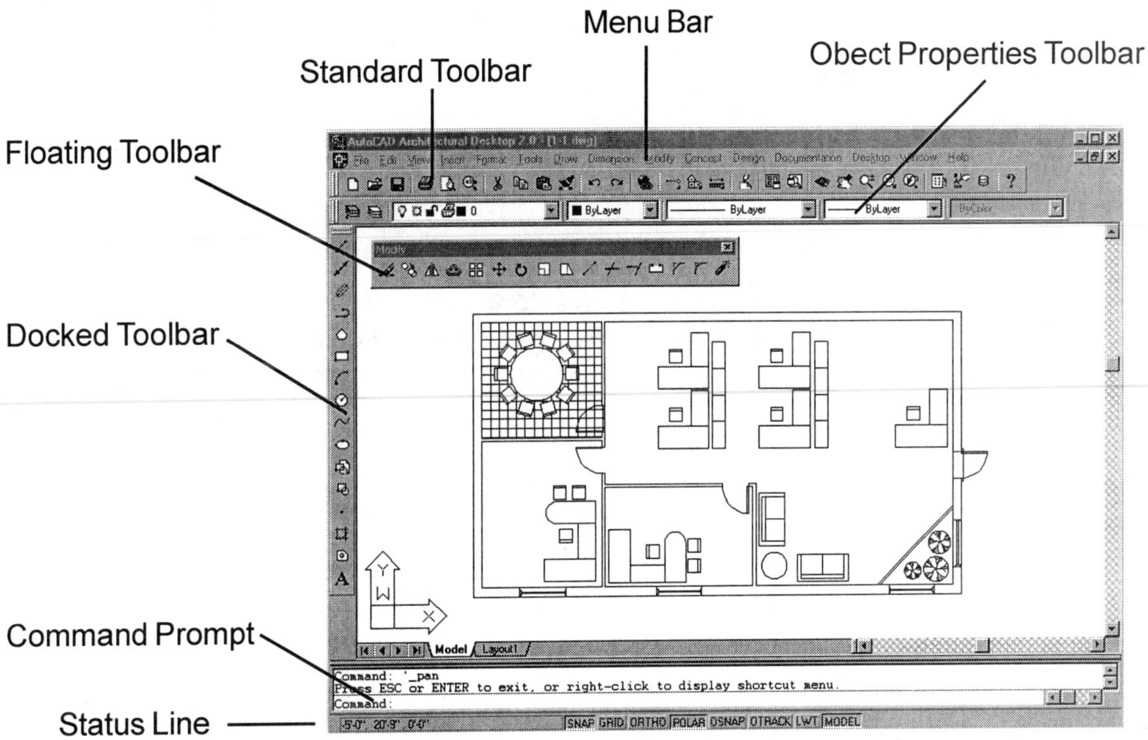

Fig. 1-1 AutoCAD Screen Display

Drawing Limits/Extents

Instead of drawing on paper, you draw within the drawing **limits** (Fig. 1-2) . The drawing limits can be thought of as the paper size. Just as you would not want to draw off of the paper and ruin your drafting table top, you do not want to draw outside of the drawing limits. While a conventional draftsperson cannot change the size of the paper after starting the drawing, the AutoCAD user can change the drawing limits at any time.

The drawing **extents** is the actual space your drawing occupies within the drawing limits. The smallest rectangle that can surround your drawing defines the drawing extents.

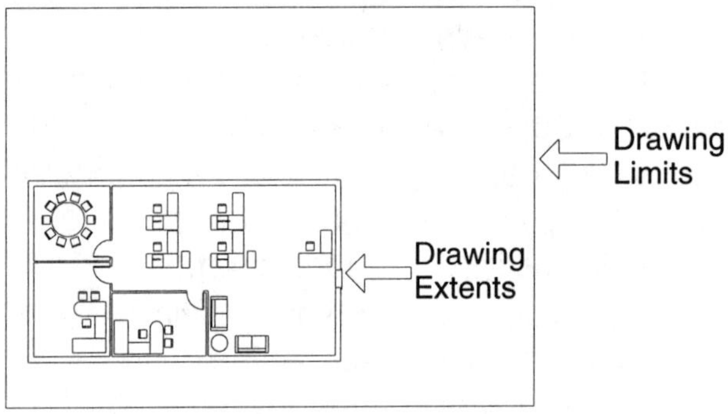

Fig. 1-2 Drawing Limits/Extents.

Drawing Display

Instead of drawing on paper, you will draw on the display screen. The screen can display the drawing to the drawing limits, extents, or parts of a drawing (Fig 1-3). In order to go from the drawing limits to just the extents, you will zoom in or use the ZOOM command. Zooming in is similar to a camera or the eye which moves closer or zooms in to view parts of the drawing. You can zoom in to any part of the drawing down to the smallest details. You can also move the drawing over just as you might move your drafting instruments to a different part of the drawing. To move over (but in the same zoom or magnification), you use the PAN command.

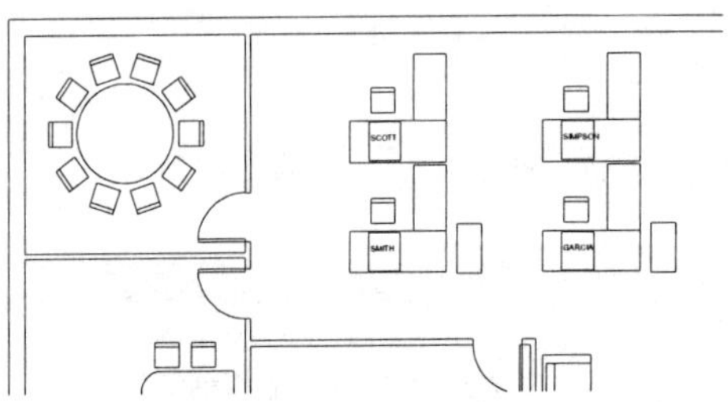

Fig. 1-3 Zoomed in to part of a drawing.

Coordinate Grid

AutoCAD uses the coordinate grid or stores drawings by means of vectors. All straight lines are drawn by locating the first coordinate location (X,Y) and then drawing a line or vector to the second coordinate location (X,Y). The illustration (Fig. 1-4) is an example of how part of a floor plan may be stored on a coordinate grid.

When you execute the LINE command for instance, AutoCAD simply prompts for the first point (or coordinate pair) and then second, and so forth. While you easily move the pointing device around, AutoCAD is storing all the information about where the line starts and ends in the database or drawing file. If you were to move the line, AutoCAD will automatically update the new location of the coordinate pairs.

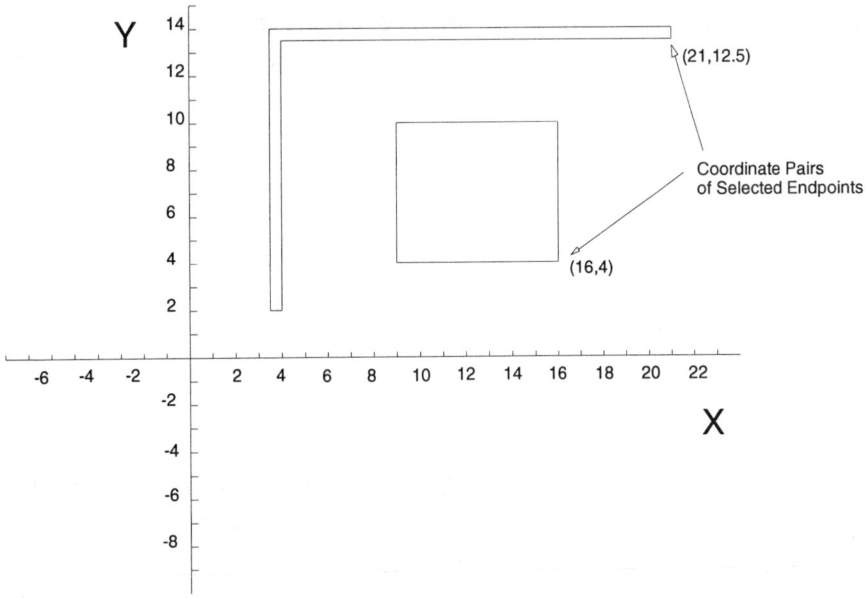

Fig. 1-4 Coordinate grid.

World Coordinate System/User Coordinate System

The X-Y coordinate grid is used to create drawings that are two-dimensional such as floor plans. When you draw on the X-Y plane, you are drawing on the **World Coordinate System** or **WCS** which is indicated by the UCS icon in the lower left corner of the screen (Fig. 1-5). To create three-dimensional images, AutoCAD uses the Z coordinate as an extrusion or what the designer would normally consider the height. The figure illustrates how the floor plan might appear in three dimensions (Fig. 1-5). Unless you specify differently, AutoCAD places entities parallel to this plane. To draw on an angle such as a roofline of a house, AutoCAD allows you to specify a **User Coordinate System** or **UCS** which essentially changes the drawing plane to what you specify. The X and Y plane then may be parallel to a diagonal plane as again in the example of the partial floor plan. To help you know the orientation of the UCS and keep your bearings in the three-dimensional world, AutoCAD provides the coordinate system icon or **UCS icon**, which are two arrows indicating the positive direction of the X and Y axes. As you can see in the illustration (Fig. 1-6) the UCS icon is aligned to the diagonal plane.

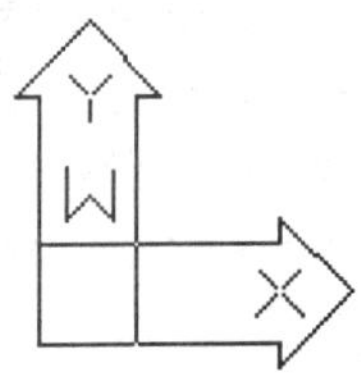

Fig. 1-5 WCS (UCS icon).

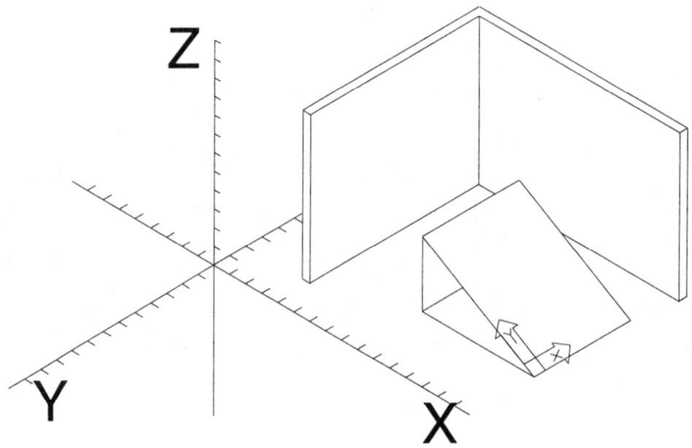

Fig. 1-6 User Coordinate system for a plane on angle.

Model/Paper Space

Much of your work in this tutorial will be done in model space where you work in full scale, i.e. an inch on the screen is equal to an inch. You can plot these drawings out by specifying the smaller scale, for instance 1/4"=1'. However, paper space allows you to plot out several drawings at different scales or views such as a floor plan at 1/4"=1' scale, elevation at 1/2"=1' scale, and section at 1"=1' scale. You accomplish this by first creating a piece of paper at full scale, then opening viewports into the model space and then establishing the scale. The paper space icon in the lower left of the screen indicate that paper space is active (Fig. 1-7).

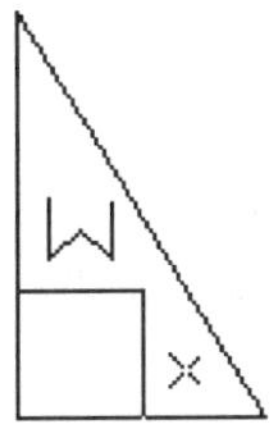

Fig. 1-7 Paper space icon.

Drawing Units

Because AutoCAD is a generic CAD software package, it provides several types of measurement or ways to describe the distance between any two coordinate pairs. The typical measurement in environmental design is feet and inches which AutoCAD describes as "architectural" drawing units. You will be using this type of units exclusively in this tutor.

Keeping Track of Your Drawing

While you are drawing, AutoCAD is updating the screen display. You must "save" your work in order to create a **drawing file** to be stored on disk. The drawing file can contain several drawings and is actually a database, meaning it is a collection of information about your drawing. The drawing file has a file name. AutoCAD drawing files have a "dwg" file extension or filetype, for example, "first.dwg." In Windows 95 you see the first part of the file name---the extension tells Windows 95 it is an AutoCAD drawing. After you are finished drawing, you end the drawing session by exiting the Drawing Editor or AutoCAD.

TUTORIAL 1

FILE MANAGEMENT BASICS

Each microcomputer is equipped with written instructions to familiarize you with the general workings of the computer and use of the keyboard. If you have not already done so, use these instructions to understand the setup of your machine, especially the operation of the keyboard.

In this tutorial, instructions are indented and highlighted in bold while the paragraphs contain explanatory information. You should not skip this explanatory information and just follow the instructions.

When drafting with AutoCAD, you create a drawing file. This file can be stored for later retrieval so that you can edit or change it. It can be copied to create backup files in case you corrupt the original. The easiest way to do these procedures is with Windows 98.

> **1. Click on the "My Computer" (or whatever the computer is named) icon.**

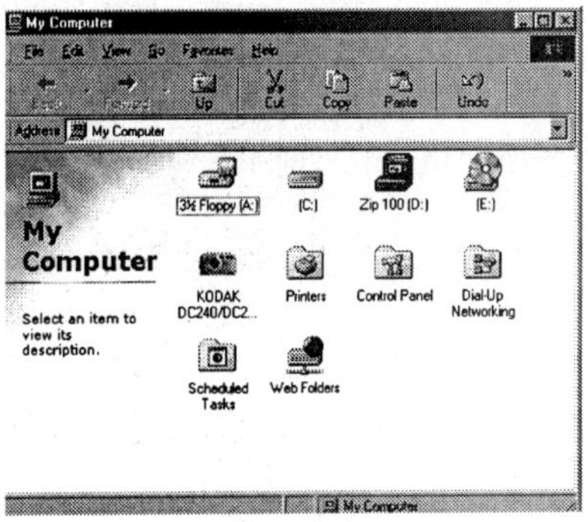

Fig. 1-6 Window of My Computer.

This window lists the various drives for this computer which may differ from yours but is typical of a stand-alone computer that is not on a network. Most computers will have an A drive for a 3-1/2" floppy diskette which is illustrated first which would be followed by an additional diskette drive B if you had one. The C drive is the hard drive containing all the programs. The D drive here is a Zip 100MB drive. The E drive is easy to recognize as a CD-ROM drive. We will not use the other folders in this tutorial.

> **1. Place a diskette in drive A and click twice on the icon above "3-1/2" Floppy (A:) drive.**

If the diskette was not formatted, you would have been asked whether you wanted to format. You also could format it by just highlighting that icon and clicking on "File" and "Format" in the window. Right now, there are no files on the diskette so it should be blank.

Displaying Files in a Window

1. **Click twice on the C drive icon to open up a window.**
2. **Click twice on Program Files folder.**
3. **Click twice on the AutoCAD R14 folder.**
4. **Click twice on the Sample folder.**

The illustration shows the sample files (Fig. 1-9).

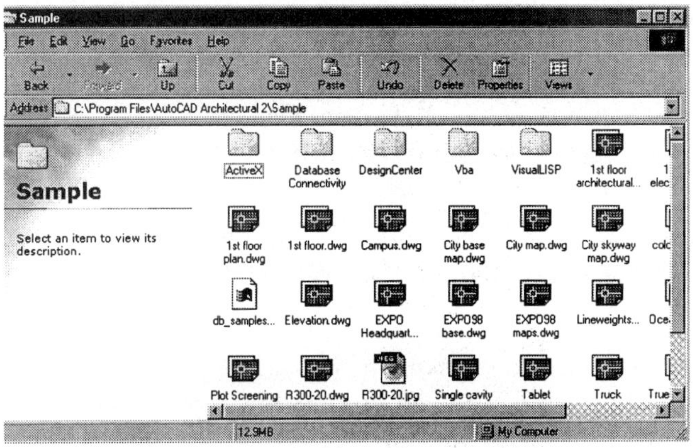

Fig. 1-9 Window of AutoCAD R14 Sample files.

Take a look at the upper right corner of this window (Fig. 1-10). You use these to minimize (make a small window at the bottom of the screen), maximize (make the window fill your screen), or close the window.

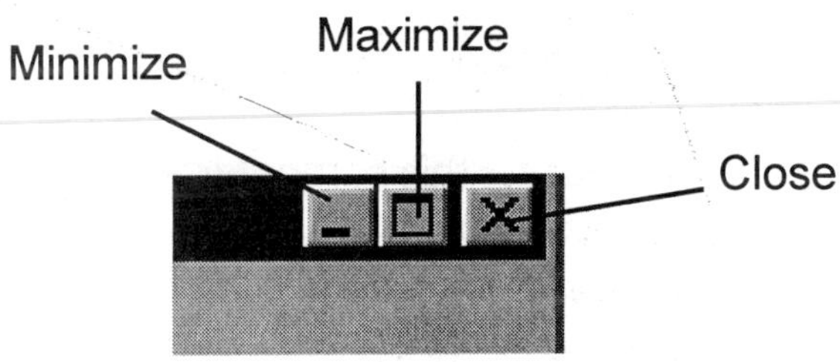

Fig. 1-10 Controlling size of window.

1. **Close all the other windows except the Sample folder and the A drive.**

You can enlarge the window to contain all the files by moving the arrow to the perimeter of the window until the double arrow appears and then clicking and dragging the window in one direction.

The Sample folder window is displaying large icons. This does not tell you much about the files. Windows 98 allows you to display your files in various ways.

1. Click on "View" at the top of the window.
2. Click on "Details."

In the illustration (Fig. 1-11), you see more information about the files including their type, size, and date of last save. You can also sort the files by date, size, or type by clicing on View and Sort. The size of the file is useful when you need to know if you have enough room on the diskette to save the file. A high density diskette holds 1.4MB (megabyes) or 1,400,000 bytes of information. The "chevy" file would not be able to be stored on a floppy diskette because its size is 2.1MB Diskettes are best used to store extra copies of drawing files. It is always risky to try to work or open a file in AutoCAD on a diskette. When AutoCAD works, it creates temporary files as well as a backup file and places them in the area where you are working. If AutoCAD runs out of space, it usually quits and sometimes will not save the file or perhaps even corrupts a file. It is best to work on the hard drive and then save a copy to the diskette at the end of your drawing session.

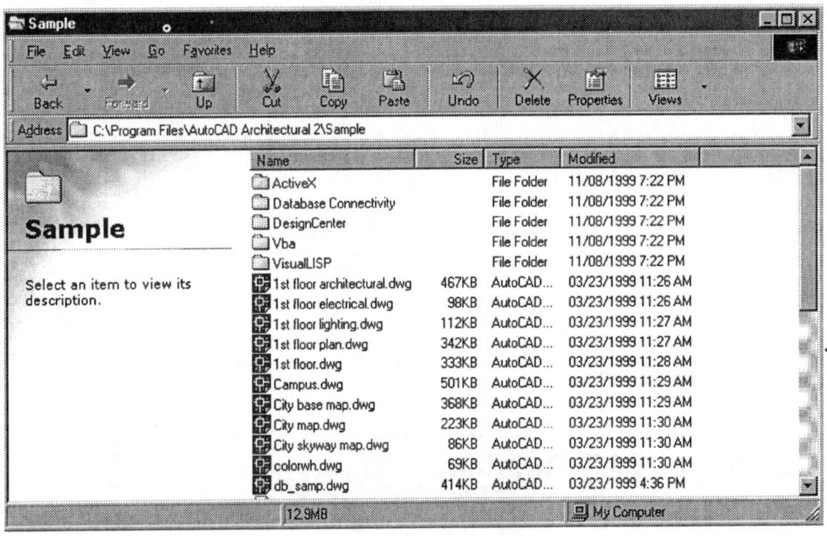

Fig. 1-11 Window listing details on files.

Copying Files

Copying a file from the hard drive to the A drive is easy; you just drag the icon to the folder:

> **1. Click on one of the small sample file icons, highlight it, and then drag to the A drive window.**

The icon should appear in the A drive window .

Had you just dragged the icon from one folder to another on the hard drive, you would have just moved the file--not copied it. The details of the file should match exactly what was in the Sample folder. If the computer had trouble copying, the size might not match which means it was not a successful copy. You might need to put another diskette in the computer and try again.

Deleting Files

Deleting files is as simple as dragging the icon to the Recycle Bin.

> **1. Drag the icon to the Recycle Bin.**

You can select a number of files for copying or deletion by making a window crossing the files. Or, you can select a whole folder to move, copy, or delete, by clicking on the folder once to highlight it.

Making Folders

You will find it useful to make a folder to organize files on your hard drive or diskette. A folder is similar to a file folder in a file cabinet because it contains several documents or files. You can make a folder for your tutorial files on your floppy diskette by:

> **1. In the A drive window, click on "File," New", and Folder.**
> **2. Click twice on the text name to change the name of the folder from "New Folder" to "CAD Tutorials."**

This has given you a brief introduction to file management. The important goal is to know where your files are and to keep current copies of your work in separate places.

Part Two: Beginning AutoCAD

TUTORIAL 2

Your First Drawing

Commands Learned:	UNITS	ERASE	PAN	ELLIPSE
	SAVE	COORDS	COPY	RECTANG
	LIMITS	SNAP	MOVE	POLYGON
	GRID	CIRCLE	U	HELP
	ZOOM	REDRAW	UNDO	
	LINE	REGEN	REDO	
	ORTHO	ARC	OOPS	

This tutorial assumes you have already loaded AutoCAD and are in the Drawing Editor. As you learned before, there are a variety of ways to enter commands or draw in AutoCAD. You can enter commands by using:

- **Icons** on the toolbars.
- **Typing** at the command prompt.
- **Menu bar** at the top of the screen.

Two other ways to enter commands are by bringing up the right screen text menu or by entering commands from a tablet menu attached to a digitizer. Neither of these methods will be used in this tutorial, but you should be aware that they exist.

This tutor will list both the command prompt method and the related icons. In some instances, you will use the menu bar to enter commands. If you are using a different version from AutoCAD Release 14, entering at the command prompt would be the best way to use this tutor.

While typing in commands is by far the slowest method to enter commands, it does help you to remember the names of the commands. After you are finished with this tutorial you should always strive for the fastest, most intuitive, and most productive way to enter commands--usually the icons.

Instructions for Using this Tutor

Instructions are indented and highlighted in the text similar to Tutorial 1. In command sequences, your input is highlighted in bold after the AutoCAD response. The appropriate icon may also be shown right above the command. When instructions are in parentheses, that means you are to carry out the instruction such as click on an object rather than typing anything. For instance:

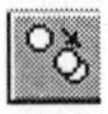

Command: **COPY**
Select objects: **(click on the circle)**

After you entered the command by clicking on the icon or typing “COPY”, you select a a circle by clicking on it. Once you enter the command, AutoCAD will prompt you for additional data or information. At these data prompts, you must enter the appropriate data such as a point or option or else AutoCAD will respond with an “invalid” message or give you further instructions.

When the Tutorial instructs you to use the menu bar, it will appear:

1. Menu Bar: Format > Units...

This instructs you to click on “Format” on the Menu Bar located at the top of the Drawing Editor, then “Units” which will bring up a dialogue box.

UNITS Command

The first task is to set up your drawing. AutoCAD sets up the default drawing units as decimal. You will change it to architectural (English units) or feet and inches with the UNITS command. This command is easiest to use by using the menu bar:

1. **Menu Bar: Format > Units...**
2. **In Type area, click on “Architectural” units.**
3. **Click on “OK” to close the window.**

LIMITS Command

Think of the drawing limits as the paper size. You will be setting the paper size in modeling space (later you will learn about another space to draw on — paper space). AutoCAD will first ask for the coordinates of the lower left corner of the screen. In most cases, you will want to keep the default value or 0,0. The upper right set of coordinates will determine the limits.

1. Menu Bar: Format > Units...

AutoCAD will respond at the command prompt:

Reset Model space limits:
Specify lower left corner or [ON/OFF] <0’-0”,0’-0”>: **(press ENTER)**
Specify upper right corner <1'-0”,0’-9">: **22',17'**

You used the ENTER key to accept the default of 0,0 for the lower left corner. The ENTER key sends the information to the computer so you always use this key after you type something in at the command or data prompt. Next you entered 22', 17' for the upper right corner of the limits which sets up the drawing "paper" for 8-1/2" x 11" paper (A size) with a 1/2"=1' scale if you were to print it out. For 1/2"=1' scale, you multiply 2 (the scale) times 11 (size) for the X direction or width to get 22 and 2 x 8-1/2 to get 17 for height. Listed below are other common limits and their corresponding scale and paper size.

Limits	Scale and paper size
144',96'	1/4" scale, 24" x 36" (D size)
288',192'	1/8" scale, 24" x 36" (D size)
96',73'	1/4" scale, 18" x 24" (C size)
192',144'	1/8" scale, 18" x 24" (C size)

In AutoCAD modeling space, you draft at full scale and print or plot out at whatever scale you choose. That is, a line that is 3 feet wide is drawn by measuring a 3-foot distance across the screen. It is easier to see how this works if you use a grid of dots to guide you.

SAVE Command

All your work is saved in RAM (Random Access Memory), which is the computer's temporary memory. If the power were to turn off, you would lose this work. Therefore, you will save your file to the disk where it will be safe:

1. Menu Bar: File > Save...
2. Click on file box and type "tut2yourinitials" in File box.
3. Click on "Save" to exit.

Your drawing has been saved in the current directory (the directory that you loaded AutoCAD from). Now save again using the icon:

1. Click on the "Save" icon .

AutoCAD does a "quick save" (QSAVE Command) and automatically saves the file. AutoCAD uses the same file name restrictions as the disk operating system or Windows . Certain characters such as the $ cannot be used. AutoCAD will tell you if you have typed an invalid filename. AutoCAD adds the file extension "dwg" automatically to the name so "tut2yourintials.dwg" is the actual file name for this file. The second time you enter a SAVE command, AutoCAD moves the previous saved file to a backup file with extension "bak" such as "tut2yourintials.bak."

The Importance of Saving Your Drawing

Get in the habit of saving your drawing periodically, every ten minutes or so. For projects or important files, make additional copies and keep in another location.

Escape Key

Press the "Esc" key to cancel. This is necessary when you are in a command and AutoCAD will not let you out of the command. The "Esc" or escape key will bring you back to the command prompt. Also, use this key when you select objects on the screen accidentally leaving grips or little blue boxes. The "Esc" key erases these grips which you learn more about later.

You will now turn off "object snap" (a feature you wil learn about in a later tutorial):

1. If OSNAP is recessed on status line, click until it is not recessed or forward to turn it off.

GRID Command

You will make a grid of dots to use as a visual reference.

Command: **GRID**
Specify grid spacing(X) or [ON/OFF/Snap/Aspect] <0'-0">: **1'**

You cannot see the grid yet because you are looking at a small area of the paper (9" x 12" or the size of the screen) so you will use the ZOOM to see the whole paper.

ZOOM Command

The ZOOM command allows you to zoom the drawing. It works much like the zoom lens on a camera. Another analogy is if you are sitting at the drafting table, you can bring your eyes close to see a portion of the drawing (zooming in) or sit farther back to see the whole drawing (zooming out). You will just enter Z for the ZOOM command--a quick step which can be done with most commonly used commands:

Command: **Z**
ZOOM
Specify corner of window, enter a scale factor (nX or XP), or
[All/Center/Dynamic/Extents/Previous/Scale/Window]<real time>: **A**

Just as you only needed to type the first letter of the ZOOM command, you can enter just the first letter of the options for the ZOOM command. Thus, you need only

type what is capitalized in the options to save time. This is true of all AutoCAD commands so never key in more than you have to in order to save time.
You should see a grid of dots. As you specified, the distance between each dot is one foot. The drawing area should look like a letter-size paper arranged horizontally.

You can turn the grid off by pressing the F7 function key or clicking GRID on the status bar. The F7 or GRID function key is one of several function keys. These keys are toggle keys, meaning you press the key once to turn something on and then press again to turn something off. You will learn more about the other function keys later in this exercise.

Finally, you are ready to draw!

LINE Command

In the following sequence, point in the approximate location that is illustrated (Fig. 2-1). The little x's or blip marks show where to point. The LINE command is found on the draw tool bar and is represented by the following icon. Either click on the icon or just type L to the command prompt as below:

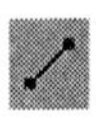

Command: **L**
LINE
Specify first point: **(click at about point A)**
Specify next point or [Undo]: **(point B)**
Specify next point or [Undo]: **(point C)**
Specify next point or [Undo]:: **(press ENTER)**

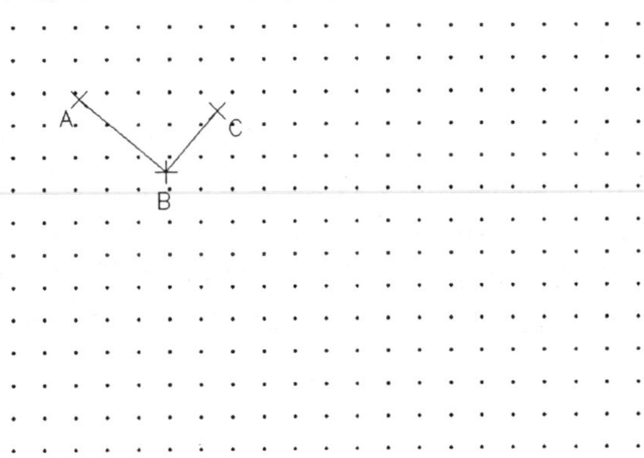

Fig. 2-1 Place lines in upper left of screen display.

You pressed the ENTER key to terminate the line sequence. When the command prompt appears, press the ENTER key again:

Command: **(press ENTER)**

This brings up the LINE command again and is a good way to repeat commands for speed. Pressing the space bar when at command prompt also will repeat a command. Next, you wil use a few connected lines and then use the "u" to undo the last line drawn (Fig. 2-2):

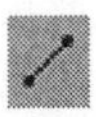

Command: **L**
LINE Specify first point: **(point A)**
Specify next point or [Undo]: **(point B)**
Specify next point or [Undo]: **(point C)**
Specify next point or [Undo]: **(point D)**
Specify next point or [Undo]: **(point E)**
Specify next point or [Undo]: **U**
Specify next point or [Undo]: **(press ENTER or right click)**

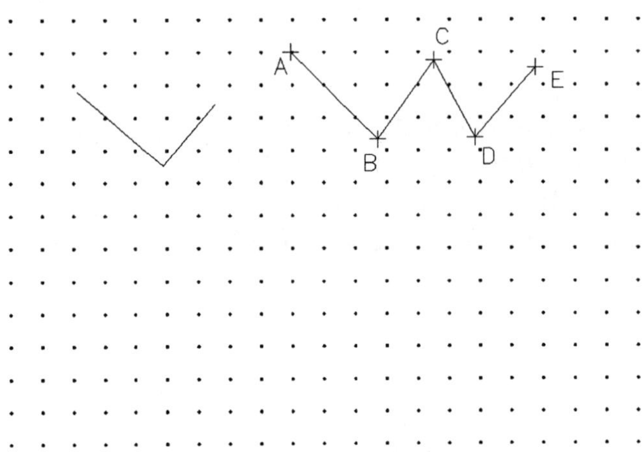

Fig. 2-2 Undoing the last line drawn.

The last line should be undone or erased. By typing U repeatedly to the prompt, you could undo back to the starting point.

ORTHO Command

When you turn ORTHO on, you draw either horizontally or vertically, much like using a T-square and triangle. While you can type ORTHO at the command prompt, it is much easier to use the function keys to toggle the command on or off. When you turn it on, ORTHO should appear on the status line at bottom of the screen.

1. Click on the ORTHO at the bottom of the screen or status line until the icon is recessed which turns it on (or press F8 key to turn ORTHO on).

Then enter draw a few lines in the approximate location as shown (Fig. 2-3):

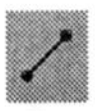

Command: **L**
LINE Specify first point: **(point A)**
Specify next point or [Undo]: **(point B)**
Specify next point or [Undo]: **(point C)**
Specify next point or [Undo]: **(point D)**
Specify next point or [Undo]: **(point E)**
Specify next point or [Undo]: **(point F)**
Specify next point or [Undo]: **(press ENTER or right click)**

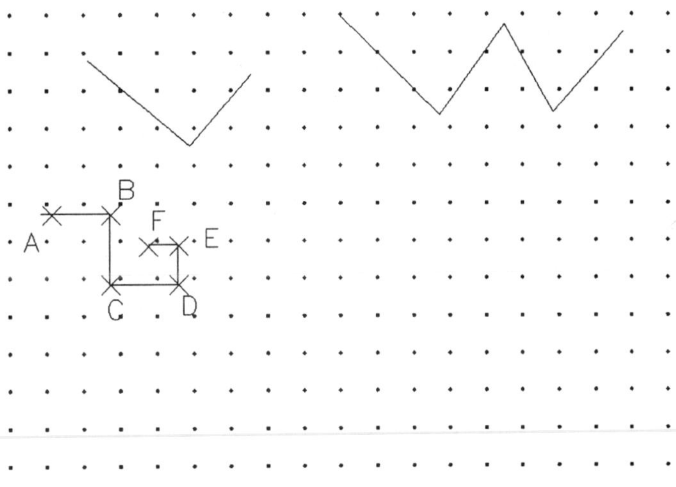

Fig. 2-3 Drawing with ORTHO on.

By now, you may have wanted to erase something you have done so the next section will give a brief introduction to edit commands.

1. Save your file by clicking on 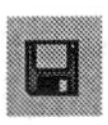 **save file icon.**

Edit Commands and Selection Sets

The edit commands allow you to modify your entities or objects. After an edit command is entered, AutoCAD asks you first to select the object to be edited with the prompt "Select objects:" You can select more than one object and these become a part of the selection set. The following are the different options or methods of designating the selection set, listed for your reference only—do not try to enter these:

ALL:	selects all objects on screen.
pointing:	point at an object with the pointing device.
Window:	place a window that completely surrounds a group of objects.
Crossing:	place a window which surround and crosses a group of objects.
Last:	the last object drawn.
Undo:	removes the last object added to the selection set.
Previous:	uses selection set created with SELECT command.
Remove:	switches from adding to removing objects from the selection set.
Add:	switches back to adding objects to the selection set.
WPolygon:	place a polygon window that completely surrounds objects.
CPolygon:	place a polygon window that crosses objects.
Fence:	slects crossing a fence created by as series of points.
BOX:	identical to window and crossing. If you make a box from left to right, it's a window and from right to left is the crossing option as above.
AUto:	automatic selection. You can choose either by pointing at an object or making a window depending on where you place crosshairs. In an empty space, it's a window.
SIngle:	instructs AutoCAD to stop the selection after the first selection for all subsequent edits.
Multiple:	instructs AutoCAD to scan the drawing just once for a group of points.
Previous:	uses the previous selection set for the edit command.

As with all options, type only what is capitalized. Pressing the Esc key will cancel you out of selecting.

ERASE Command

You will practice a few of the selection set options with the ERASE command. The ERASE icon is on the modify toolbar. If you were not sure that was the erase icon, look on the status line at the bottom while you have the pointer over the icon. It says "Removes objects from a drawing; erase"-- a good way to hunt for the right icons to use. Now click on the erase icon or enter E to use the ERASE command:

Command: **E**
ERASE
Select objects: **L**
Select objects: **(press ENTER)**

Notice that the object became "dashed" to indicate it had been selected and AutoCAD responded with "1 found"—it is important to notice what AutoCAD found. When you pressed the ENTER key, AutoCAD knows you do not want to select any more objects, so it erases the object selected to finish the command.

Command: **(press ENTER)**
ERASE
Select objects: **(just point on one of the lines with the pointing device)**

In this instance, you just point at an object. When AutoCAD responds, type "U" to undo the last selection:

Select objects: **U**
Select objects: **(select a line)**
Select objects: **(press ENTER)**

COORDS Command

The COORDS command turns on coordinate tracking which tracks the X-Y coordinate location of the crosshairs.

1. Press F6 key twice to enable coordinate tracking.

SNAP Command

The SNAP command will snap the crosshair to designated spacing. Enter:

Command: **SNAP**
Specify snap spacing or [ON/OFF/Aspect/Rotate/Style] <0'-0">: **1'**

1. Move the crosshairs to see that it snaps to points on the grid.

Change snap to 6" by:

Command: **SNAP**
Specify snap spacing or [ON/OFF/Aspect/Rotate/Style] <0'-0">: **6**

1. **Move the crosshairs to see that it snaps to points within and on grid.**
2. **Turn snap off by clicking on SNAP on status bar or use F9 key.**
3. **Move the crosshairs to see that SNAP is off.**
4. **Turn snap on again by clicking on SNAP on status bar or use F9 key.**

Note you did not have to put the inches sign (") because the inch increment is the default increment, meaning AutoCAD assumes you mean inches when you do not enter the feet sign (').

Drawing a Square by Snapping to Points

You will draw a square by using SNAP, ORTHO, and following coordinate tracking as well as by using the "C" option of LINE command which means closing a polygon (Fig. 2-4):

1. **Check on the status line to make sure ORTHO and SNAP are on.**

Command: **L**
LINE Specify fist point: **(point A in approximate location as Fig. 2-4)**
Specify next point or [Undo]: **(point B or when coordinate tracking indicates**
3'- 0"< 0)
Specify next point or [Undo]: **(point C or when coordinate tracking indicates**
3'- 0"< 270)
Specify next point or [Undo]: **(point D or when coordinate tracking indicates**
3'- 0"< 180)
Specify next point or [Undo]: **C**

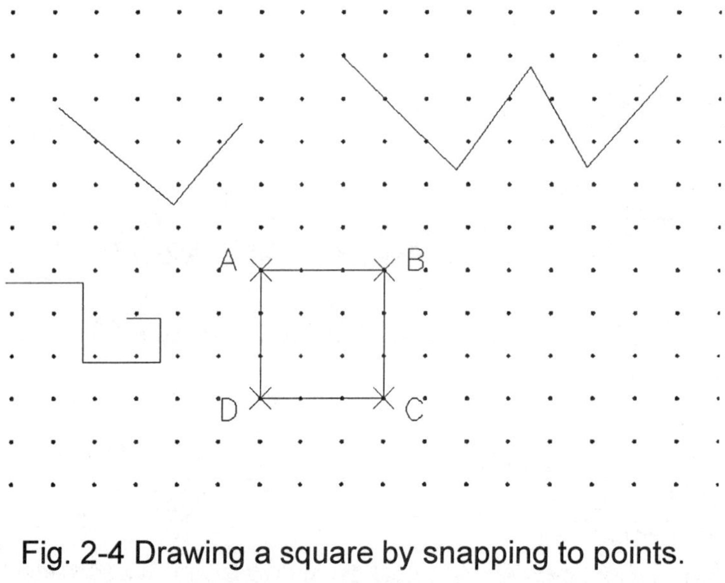

Fig. 2-4 Drawing a square by snapping to points.

Drawing a Square by Entering Polar Coordinates

You noticed that coordinate tracking indicated angles such as "3'<270" meaning AutoCAD was locating a point for the line 3 feet in the 270 degree direction. AutoCAD measures angles counterclockwise as in the illustration (Fig. 2-5).

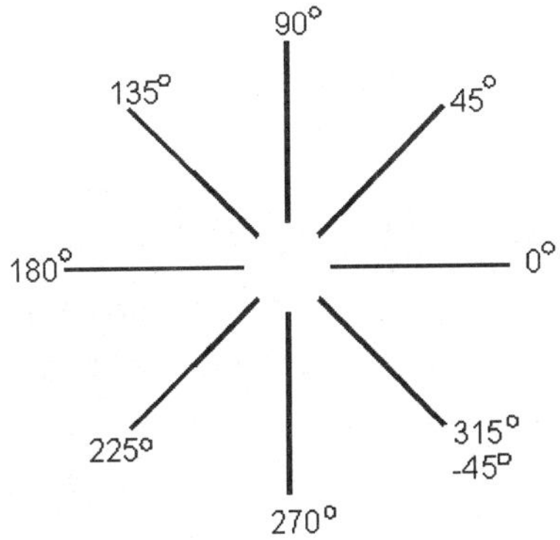

Fig. 2-5 Polar coordinates.

Next, you will draw a rectangle (Fig. 2-6). First, you will tell AutoCAD you want to start the line at X=1' and Y=2' or at X-Y coordinate pair location (1',2'). On the subsequent point, you enter the @ sign to connect the line to the end of the last line entered.

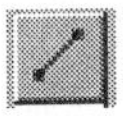

Command: **L**
LINE Specify first point: **1',2'**
Specify next point or [Undo]: **@4'1<90**
Specify next point or [Undo]: **@3'3<0**
Specify next point or [Undo]: **@4'1"<-90**
Specify next point or [Undo]: **C**

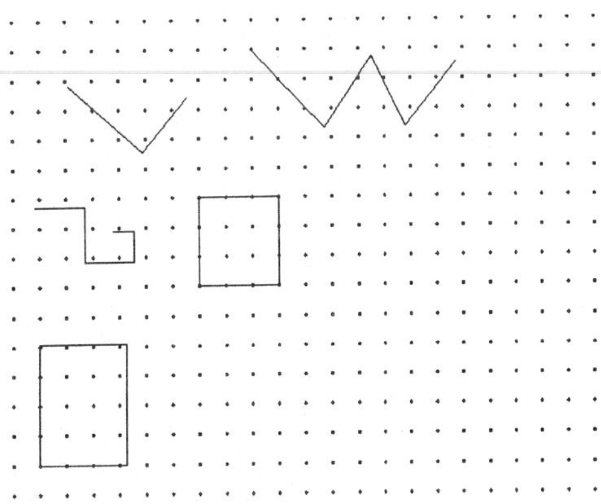

Fig. 2-6 Drawing a square by using polar coordinates.

CIRCLE Command

Circles can be made in a variety of ways. You will draw a 3' diameter circle by using SNAP. For the center point, point approximately in the same location as illustrated (Fig. 2-7).

1. Make sure SNAP is on and set at 6 inches.

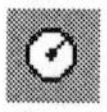

Command: **C**
CIRCLE
Specify center point for circle or [3P/2P/Ttr (tan tan radius)]: **(point A)**
Specify radius of circle or [Diameter]: **(point when coordinate tracking indicates 1'-6"< 0)**

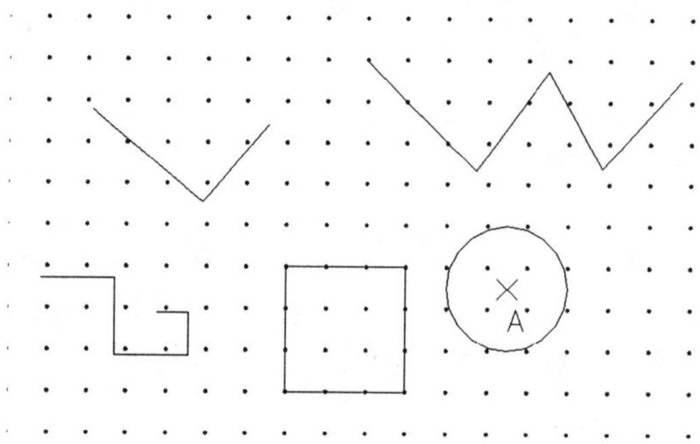

Fig. 2-7 Circle by pointing.

Now make another 3' circle by just entering the radius of 18 inches:

Command: **C**
Specify center point for circle or [3P/2P/Ttr (tan tan radius)]: **(point A)**
Specify radius of circle or [Diameter]: **18**

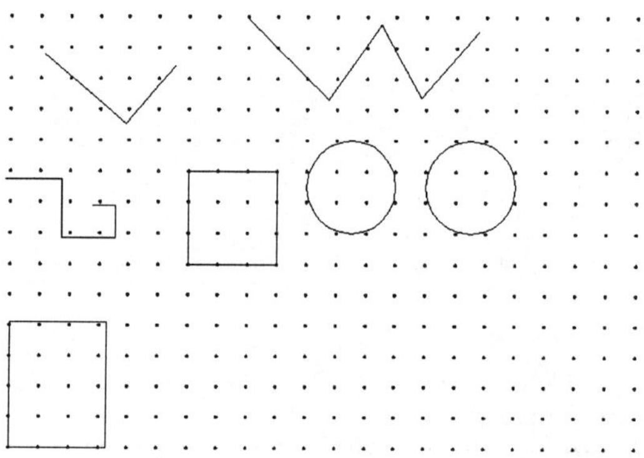

Fig. 2-8 Circle by entering radius.

Another way to draw a circle is by specifying three points on its circumference (Fig. 2-9):

Command: **C**
CIRCLE Specify center point for circle or [3P/2P/Ttr (tan tan radius)]: **3P**
Specify first point on circle: **(point A)**
Specify second point on circle: **(point B)**
Specify third point on circle::**(point C)**

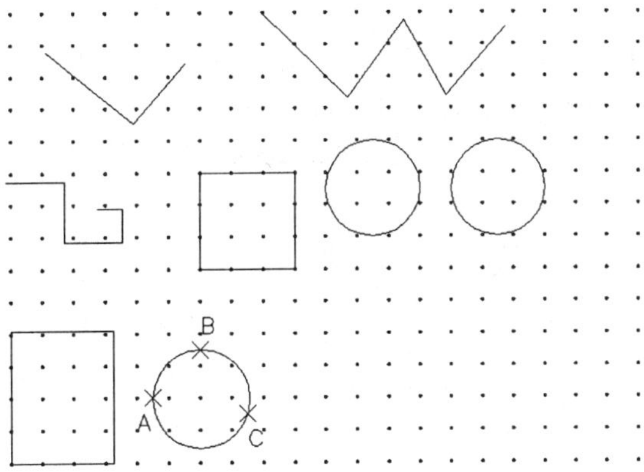

Fig. 2-9 Three-point circle.

Still another method of drawing a circle is to specify the two endpoints of the diameter Fig. 2-10):

Command: **C**
CIRCLE Specify center point for circle or [3P/2P/Ttr (tan tan radius)]: **2P**
Specify first end point of circle's diameter: **(point A)**
Specify second end point of circle's diameter:: **(point B)**

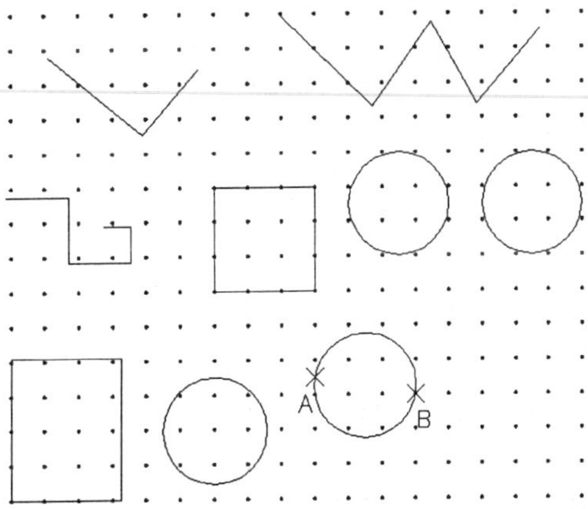

Fig. 2-10 Two-point circle.

Finally, you can draw a circle by specifying two lines and/or circles to which the circle will be drawn as tangent. You will draw this circle in the lower rectangle (Fig. 2-11):

1. Click on SNAP on the status line or press F9 key to turn Snap off.

Command: **C**
CIRCLE Specify center point for circle or [3P/2P/Ttr (tan tan radius)]:**TTR**
Specify point on object for first tangent of circle: **(point on line at about A)**
Specify point on object for second tangent of circle: **(point on line at about B)**
Specify radius of circle <5'-0 7/16">: **1'**

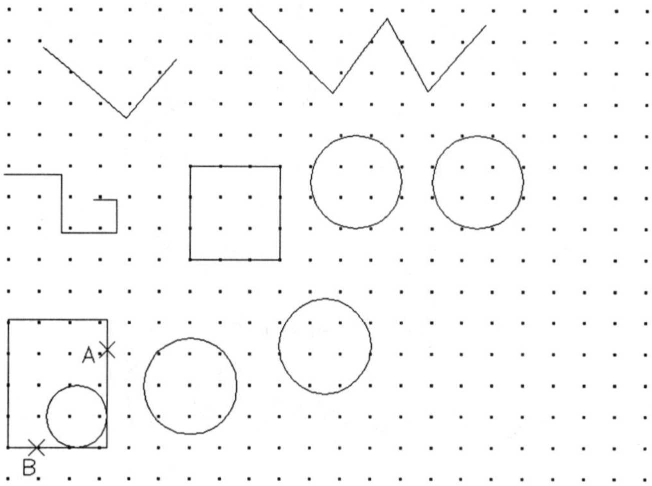

Fig. 2-11 Circle that is tangent to radius.

REDRAW, REGEN Commands

Sometimes, you will want to redraw the screen display by using the REDRAW command. You can just type R to the command prompt or click on “Redraw on “View” on the menu bar:

Command: **R**

The REGEN command will regenerate the whole drawing and thus takes longer, but this may be required after executing some commands.

The  icon redraws all the viewports---a feature you will learn about later.

ARC Command

The ARC command draws arcs in a variety of ways. First you will specify three points on the arc similar to three points on the circle. Put in the approximate size and location as in the illustration (Fig. 2-13) .

Command: **ARC**
Specify start point of arc or [CEnter]: **(point A)**
Specify second point of arc or [CEnter/ENd]: **(point B)**
Specify end point of arc: **(point C)**

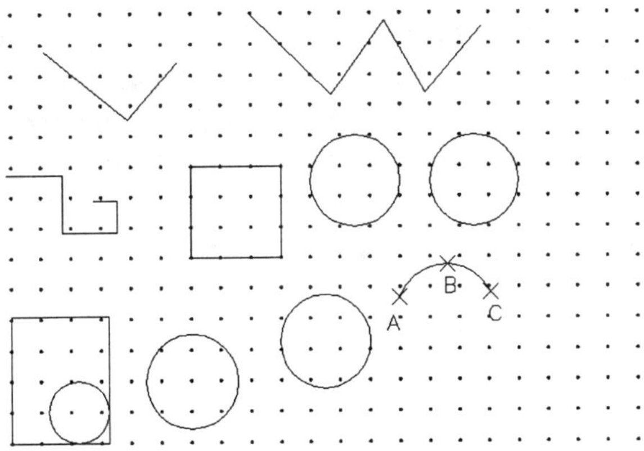

Fig. 2-13 Three points on the arc.

Another method of drawing an arc follows: (Fig. 2-14).

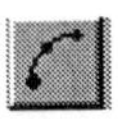

Command: **ARC**
Specify start point of arc or [CEnter]: **(point A)**
Specify second point of arc or [CEnter/ENd]: **CE**
Specify center point of arc: **(point B)**
Specify end point of arc or [Angle/chord Length]: **(point C)**

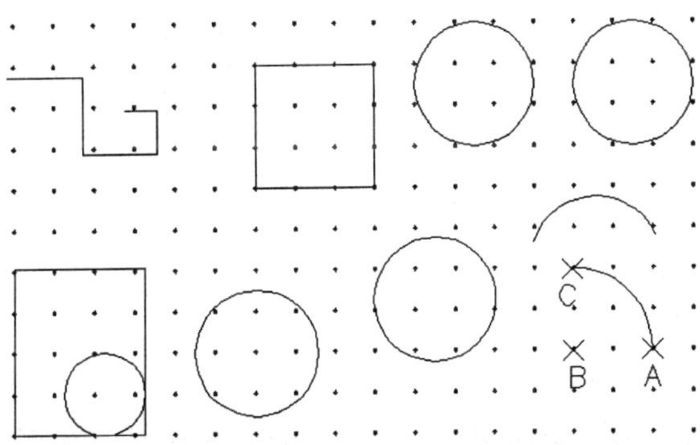

Fig. 2-14 Start-center-endpoint arc.

Notice that AutoCAD draws the arc counterclockwise just as all angles in AutoCAD are drawn in a counterclockwise direction. Now it is time to learn more about zoom.

ZOOM Command with Window Option

As mentioned before, the ZOOM command works much like a camera—to zoom in to parts of the drawing and then back out again. The illustration below shows how your screen may look. The dashed line in th eillustration is the window you make around the circle and box (Fig. 2-15):

Command: **Z**
ZOOM
Specify corner of window, enter a scale factor (nX or nXP), or
[All/Center/Dynamic/Extents/Previous/Scale/Window] <real time>: **(point on screen to make window as in illustration)**
Specify opposite corner: **(point to make window as in illustration)**

You should see the display similar to the illustration (Fig. 2-16).

Fig. 2-15 Window for zoom.

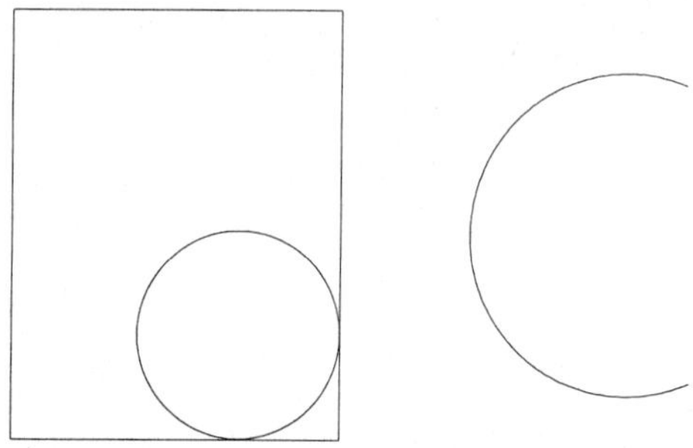

Fig. 2-16 Zoomed in area.

PAN Command

The PAN command lets you move the drawing over and is also most easily used with an icon on the standard toolbar:

1. Click on:

2. Click the hand in the middle of the screen and then move to the right.
3. Press ESC or ENTER to exit, or right-click to display shortcut menu.

The screen display moves over to the right the same distance between the first and second points. Note that your drawing is at the same magnification as before; you just moved your eye or screen display over. You also have not moved any entities they stay in the same location within the limits.

Panning with the Scroll Bars

A simple way to pan is to use the scroll bars located on the right and bottom of the graphics window.

1. Try using the scroll bars to pan.

Important ZOOM Command Options

In the previous PAN command, you panned the drawing in realtime. The ZOOM command can also be done in realtime.

1. Click on the "Zoom realtime" icon on the standard toolbar:

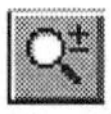

2. Move the magnifying glass.
3. Now right click on the screen--a small dialogue box comes up.
4. Click on "Zoom Original" to see the drawing as before.
5. Right click and click on "Zoom Extents."
6. Right click on "Zoom Window" to use the window option again.

Zoom previous is similar to zoom original and has its own icon on the standard toolbar:

1. Click on the "Zoom previous" icon. 

The only other zoom icon on the standard toolbar is a fly-out which contains most of the rest of the zoom options.

1. Click on the icon to the right of the zoom realtime icon.

2. Then hold the mouse button down to fly out the other icons:

all
window
dynamic
scale
center
in
out
all
extents

Your display might be different because the zoom you just used is placed at the top. You will want to practice with most of these zoom icons.

1. Use the "Zoom in"  **icon.**

If you paid attention to the command prompt, you will see all the options listed. For additional experience, the tutor will list the command sequence in addition to the icon--try entering from the command prompt on some of the zoom options. The next zoom option to try is the zoom extents which is very useful when you want to zoom a floor plan and do not want to see the whole drawing limits.

Command: **Z**
ZOOM
Specify corner of window, enter a scale factor (nX or nXP), or
[All/Center/Dynamic/Extents/Previous/Scale/Window] <real time>: **E**

Next, you will use the zoom window icon to zoom to any area of screen:

Command: **Z**
ZOOM
Specify corner of window, enter a scale factor (nX or nXP), or
[All/Center/Dynamic/Extents/Previous/Scale/Window] <real time>: **(click for lower corner of window)**
Other corner: **(point to upper right corner)**

Command: **Z**
ZOOM
Specify corner of window, enter a scale factor (nX or nXP), or
[All/Center/Dynamic/Extents/Previous/Scale/Window] <real time>: **D**

The "Dynamic" option allows you to dynamically select a window in the drawing. It sometimes saves you a step so it is worth the time to learn how to use it. A white box is around the drawing limits. A smaller white box appears as the size of the view while a green dashed window (magenta on some monitors) appears for the previous view location.

1. **Move the box to another location and point to select the center point.**
2. **Dynamically stretch the window to a size you might want.**
3. **If you point again, you can move the center point.**
4. **To finish the command, press ENTER or right click.**
5. **Click on the**  **icon to zoom all of the drawing.**

The dynamic option may take some practice, but it is well worth using on a regular basis. The ZOOM window, previous, extents, all, and dynamic options are the most commonly used. At the the end of this exercise is an additional section which has some more options for the ZOOM command.

COPY Command

A powerful feature of any CAD system is the ability to copy. Once you have selected the object or objects to copy, AutoCAD will ask for a base point or displacement. Placing this point on the object gives a reference point. The second point of displacement then establishes the distance away from the first point or where AutoCAD will place the copied object.

An analogy may help you to understand displacement. Suppose you moved a chair from one location in the room to the opposite side. When you put your hands on the chair, that is the base point or displacement. Where the chair finally ended up is the second point of displacement.

The copy icon is found on the Modify Toolbar. Follow the points in the illustration (Fig. 2-17).

1. Turn ORTHO off (F8 key or click ORTHO on status line).

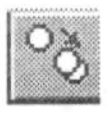

Command: **COPY**
Select objects: **(point on arc at point A)**
Select objects: **(press ENTER or right click)**
Specify base point or displacement, or [Multiple]: **(point B)**
Specify second point of displacement or <use first point as displacement>: **(point C)**

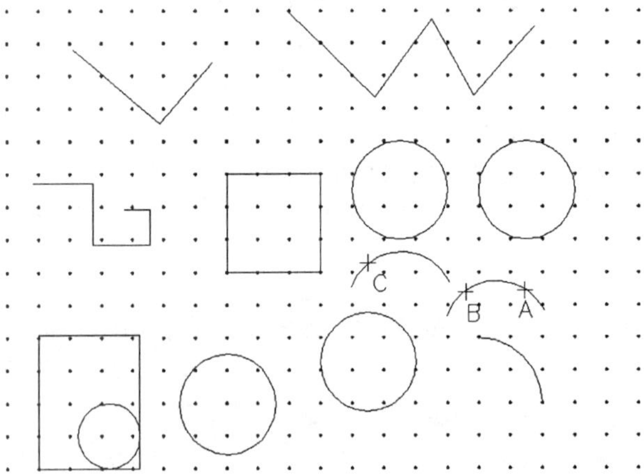

Fig. 2-17 Copying the arc.

It makes sense to point on the object for the first point of displacement because it is similar to dragging the copy to the new location.

Copying or Cutting Using the Clipboard

If you are familiar with other Windows programs, you know how to use the "Cut-Copy-Paste" icons:

Cut--Copy--Paste

Click on the icon, then AutoCAD will prompt you to select the object for copying. To paste the copy, just click on the paste icon. AutoCAD will prompt you for the insertion point or where you want to place it. This is a good technique for copying objects from one drawing to the other.

Another way to use the COPY command is to make multiple copies:

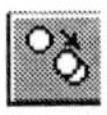

Command: **COPY**
Select objects: **(point to one of the arcs)**
Select objects: **(press ENTER or right click)**
Specify base point or displacement, or [Multiple]: **M**
Specify base point: **(point on the object)**
Specify second point of displacement or <use first point as displacement>: **(point to where you want first copy)**
Specify second point of displacement or <use first point as displacement>: **(point to where you want second)**
Specify second point of displacement or <use first point as displacement>: **(press ENTER or right click)**

You could have done more copies than two. When you press ENTER to the second point of displacement, it terminates the command.

Using Windows for Selecting Objects

Another selection method is to make a rectangular "window" around entities. The illustration shows approximately where you should make this window (Fig. 2-18):

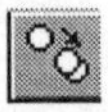

Command: **COPY**
Select objects: **(point A)**
Specify opposite corner: **(click upper right corner at point B)**
Select objects: **(press ENTER or right click)**
Specify base point or displacement, or [Multiple]: **(press Esc key to escape)**

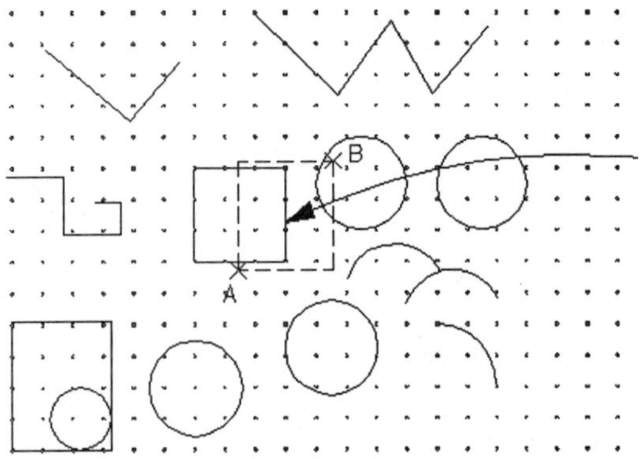

Only this line will be selected with a left to right window.

Fig. 2-18 Selecting with a window.

Next, you will use the window crossing selection option (Fig. 2-19).

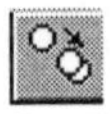

Command: **COPY**
Select objects: **(point A)**
Specify opposite corner: **(click on the lower left cornerr at point B)**
Select objects: **(press ENTER or right click)**
Specify base point or displacement, or [Multiple]: **(press Esc key to escape)**

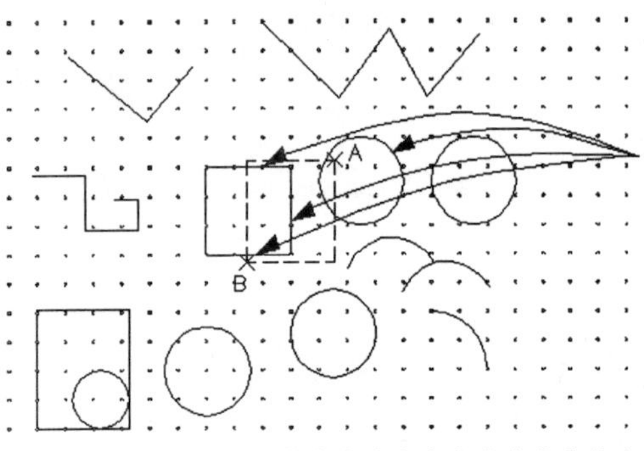

All these objects are selected with a right to left or crossing window.

Fig. 2-19 Window crossing selection

Notice that the lines that crossed the window were selected. Another feature is that the crossing window appears dashed while the enclosing window appears solid.

The “WP” option is similar to the window option. It allows you to draw a polygon around a group of objects. As with the window option, the polygon must completely surround the objects. The illustration shows how you can make a polygon to avoid selecting the large circle (Fig. 2-20).

Command: **E**
ERASE
Select objects: **WP**
First polygon point: **(point A)**
Specify endpoint of line or [Undo]: **(point B)**
Specify endpoint of line or [Undo]:**(point C)**
Specify endpoint of line or [Undo]: : **(point D)**
Specify endpoint of line or [Undo]: **(point E)**
Specify endpoint of line or [Undo]: **(press ENTER)**
Select objects: **(press ENTER or right click)**

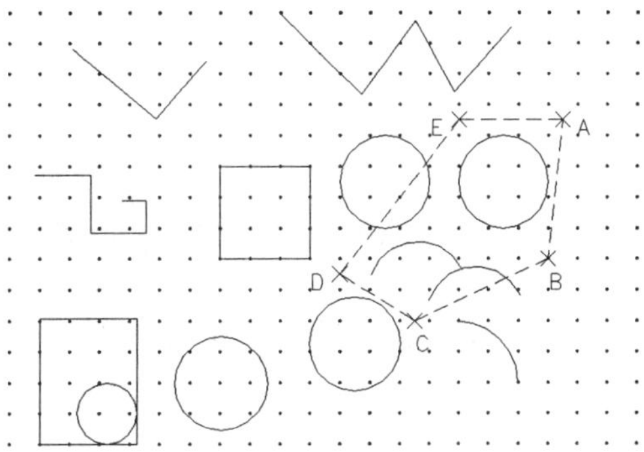

Fig. 2-20 Selection with a polygon.

The “WC” option is similar except objects that cross the polygon will also be selected.

The “Remove” allows you to remove some objects:

Command: **E**
ERASE
Select objects: **(select 3 objects with a window)**
Select objects: **R**
Remove objects: **(point to select 2 of the objects)**
Remove objects: **A**
Select objects: **(press ENTER or right click)**
Command: **OOPS**

MOVE Command

The MOVE command allows you to move an entity from its present location to a new one without changing its size or orientation. Similar to the COPY command, AutoCAD asks for points of displacement. Move the circle in the approximate location as shown (Fig. 2-21).

Command: **MOVE**
Select objects: **(point on circle at A)**
Select objects: **(press ENTER or right click)**
Specify base point or displacement: **(point B or location where you want to move from)**
Specify second point of displacement: **(point C or second location where you want to move to)**

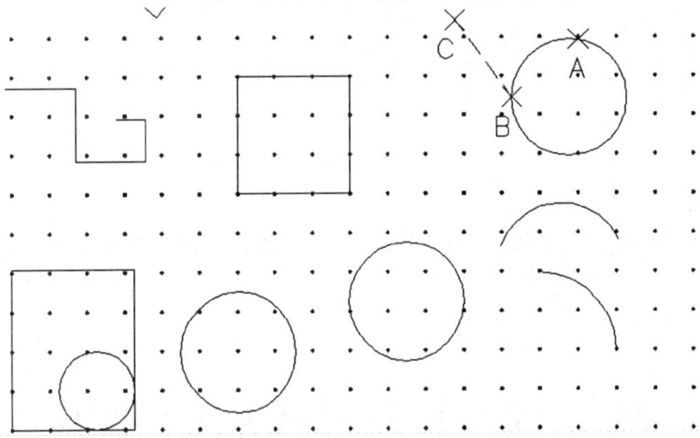

Fig. 2-21 MOVE is similar to the COPY command.

U Command

The U command allows you to undo the most recently executed command. It is found on the Standard Toolbar:

1. Draw a circle with the CIRCLE command.

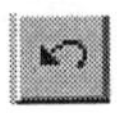

Command: **U**
CIRCLE
Command:

AutoCAD lists the last command executed which was the CIRCLE command and removes the circle.

REDO Command

AutoCAD lets you change your mind again with the REDO command. The icon is conveniently found next the undo command.

Command: **REDO**

UNDO Command

The UNDO command typed at the command prompt will allow you to undo several commands. In the next sequence, you will draw a few entities and use the UNDO command to undo several. Do not be concerned where you put them on the screen:

1. **Move the circle to a new location by using the MOVE command.**
2. **Use the COPY command to copy an object.**

You will undo the last two commands in the reverse sequence that you drew them by specifying a number:

Command: **UNDO**
Enter the number of operations to unto or [Auto/Control/Begin/End/Group/Mark/Back]<1>: **2**
COPY MOVE

UNDO Command - Mark and Back Options

You can place a marker after you enter commands to tell AutoCAD not to undo any commands before that marker but be careful—if you do not place the mark you can undo your whole drawing! This may be useful when you want to experiment. Suppose you want to save all you have done but want to play a little with some new lines. In the next sequence, you "mark" your work first, draw a few lines, then undo them back to the mark:

Command: **UNDO**
Enter the number of operations to unto or [Auto/Control/Begin/End/Group/Mark/Back]<1>: **M**

1. **Use the LINE command to draw a few lines anywhere you want.**

Now use the UNDO command with the back option:

Command: **UNDO**

Enter the number of operations to unto or [Auto/Control/Begin/End/Group/Mark/Back]<1>: **B**

The lines drawn since the marker was placed are erased. If you accidentally undid your whole drawing—do not panic. Enter a REDO command right after to restore the drawing.

OOPS Command

This command brings back the last erased object:

Command: **OOPS**

For the next three draw commands, draw the ellipses and polygons in any empty space. left.

ELLIPSE Command

You can draw an ellipse in several ways but the easiest is to specify three points as in the illustration (Fig. 2-22):

Command: **ELLIPSE**
Specify axis endpoint of ellipse or [Arc/Center]: **(point A)**
Specify other endpoint of axis: **(point B)**
Specify distance to other axis or [Rotation]: **(point C)**

The first two points specify the width or major axis of the ellipse while the last point specifies the minor axis or the length.

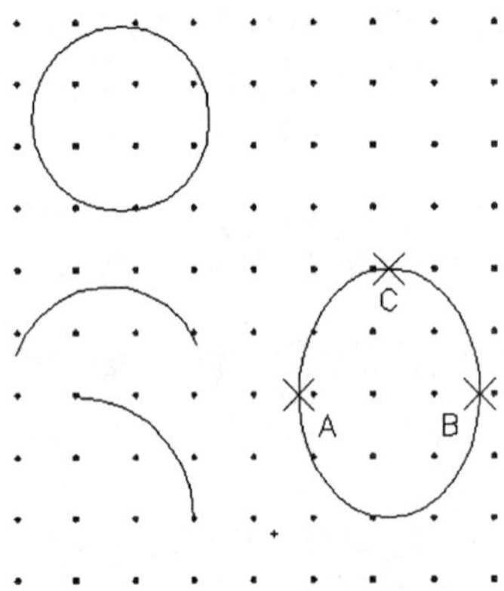

Fig. 2-22 Three points on an ellipse

RECTANG Command

The RECTANG command is most easily used by clicking on the icon on the draw toolbar:

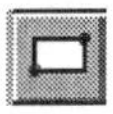

Command: **RECTANG**
Specify first corner point or [Chamfer/Elevation/Fillet/Thickness/Width/]: **(click on screen)**
Specify other corner point: **(click on other corner to make a rectangle)**

Notice some of the options for this command. Width changes the thickness of the lines while Fillet rounds the corners if you enter a radius. Chamfer will cut off the corners--you just specify the distance from each corner. Elevation and Thickness are related to the 3-D functions so do not try those for now.

1. Try making several types of rectangles using the options (Fig. 2-23).

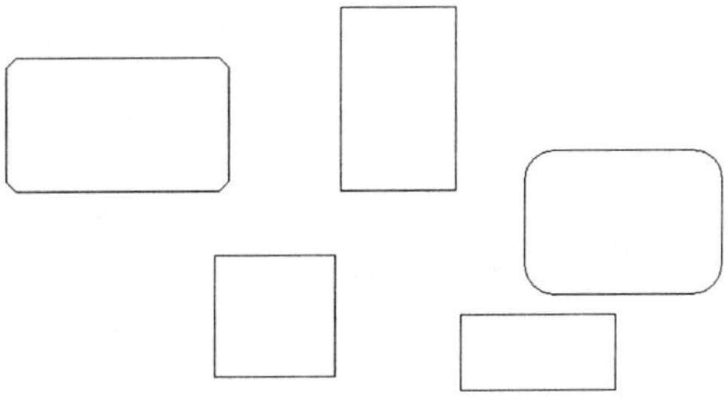

Fig. 2-23 The RECTANG command.

You can also specify the exact size of the rectangle by entering the first point and typing the distance to go in the X and Y direction:

Command: **RECTANG**
Specify first corner point or [Chamfer/Elevation/Fillet/Thickness/Width/]: **(click on screen)**
Specify other corner point: **@3',6'**

The rectangle should be 3 feet wide (3 feet in X direction) and 6' high (6 feet in Y direction).

POLYGON Command

The POLYGON command draws two-dimensional polygons with anywhere from 3 to 1024 sides. AutoCAD first asks for the number of sides and the center of the polygon. Next, AutoCAD asks whether the polygon will be inscribed within a circle, in which case the vertices of the polygon will lie on the circle, or whether the polygon will be circumscribed, in which case the midpoints of the edges are on the circle. You will draw both types (Fig. 2-24):

Command: **POLYGON**
Enter number of sides: **5**
Specify center of polygon [Edge]: **(point A)**
Enter an option [Inscribed in circle/Circumscribed about circle]<1>: **I**
Specify radius of circle: **(point B)**
Command: **POLYGON**
Number of sides: **6**
Specify center of polygon [Edge]: **(point A)**
Enter an option [Inscribed in circle/Circumscribed about circle]<1>:C
Specify radius of circle: **(point B)**

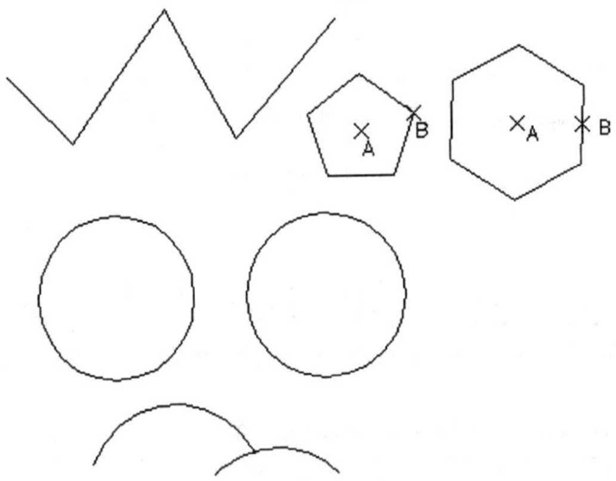

Fig. 2-24 Inscribed and circumscribed polygons.

HELP Command

This was just the first exercise and you have already learned many commands. Until you become proficient with AutoCAD, you may have trouble remembering all of the commands or what they do. AutoCAD provides an extensive help function to facilitate your learning:

1. Menu Bar: Help > Help Topics.
2. In Help Topics, click on "Contents" Tab (Fig. 2-25).

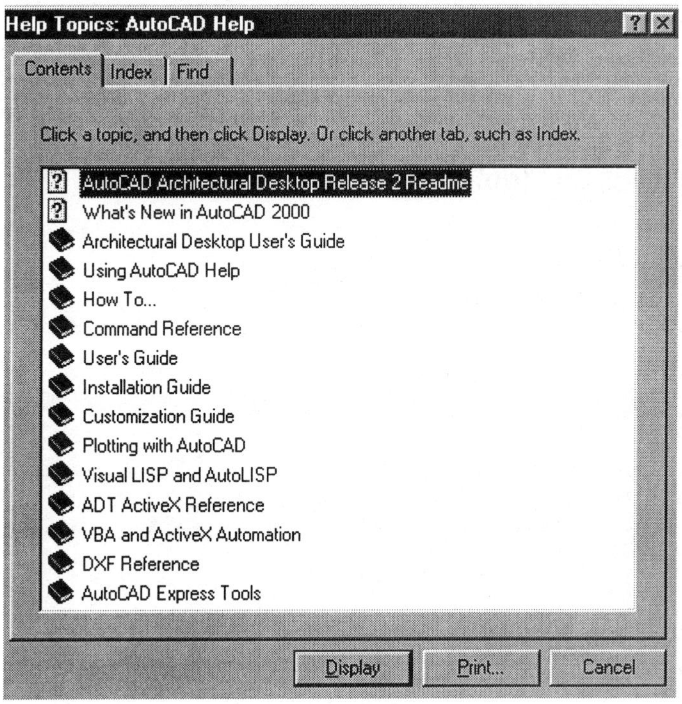

Fig. 2-25 AutoCAD Help Topics dialogue box.

3. Click on "Command Reference" and then "Command Aliases."

You already learned some of these keyboard shortcuts for entering commands (e.g. L for LINE, C for CiIRCLE). Browse through some of the others and make notes (POL for Polygon, CP for COPY, etc.).

4. Click on "Help Topics" Tab to bring back the "Help Topics" window.
4. Click on "Index" tab and enter "polygon" in the first box (Fig. 2-26).

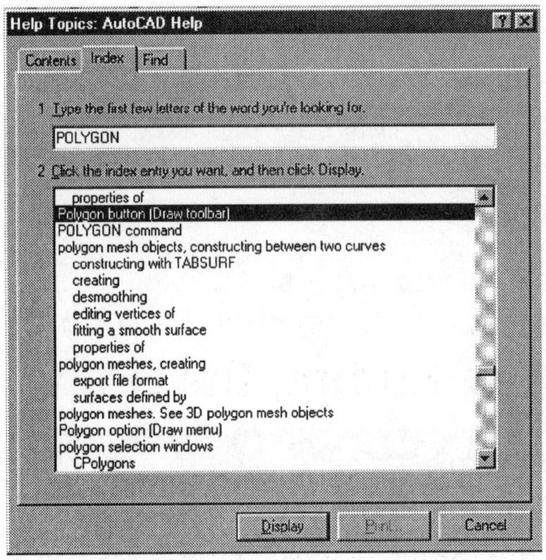

Fig. 2-26 Index of topics.

5. Highlight "Polygon button (Draw toolbar)."
6. Click on "Display" (Fig. 2-27).

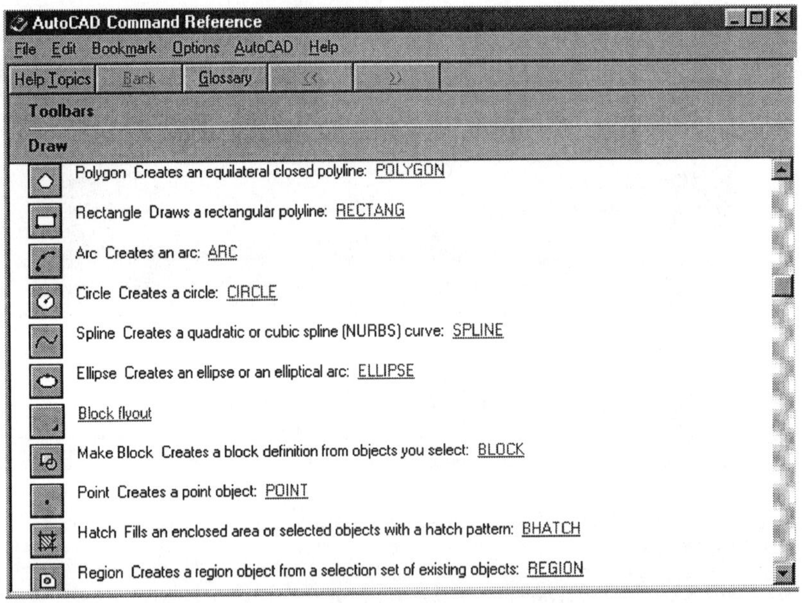

Fig. 2-27 Information about draw command icons.

You can click on the highlighted words or "hot links" to learn more about icons or buttons.

The last task is to end your drawing session or exit AutoCAD which is very important to do.

QUIT Command

You can exit AutoCAD by closing the window, typing QUIT, or clicking on "File" and "Exit" on the menu bar. If you have not saved your file, AutoCAD will ask you whether you want to save changes to your file.

1. Menu Bar: File > Exit.

The Importance of Ending the Drawing Session

While you are in the Drawing Editor, AutoCAD sets up temporary work files for its use. When you properly end the drawing session, AutoCAD removes these files. If you accidentally turn off the machine while in the Drawing Editor, you may have AutoCAD temporary files on your disk. These temporary files have a "$" in them. Do not delete these while you are in AutoCAD because they are needed. But if they are left on the disk after you exit AutoCAD, they should be deleted by use of Windows 95 after you exit AutoCAD. Another possible outcome of shutting the machine off without exiting AutoCAD is that the next time you open your file, AutoCAD may state that your file needs recovery. Enter yes to the prompts; otherwise, AutoCAD will not open the file.

SHORT ANSWER/DISCUSSION QUESTIONS

1. If you are drawing a series of connected lines and the last one you drew was incorrect, what should you respond to the "Specify next point or [Undo]:" prompt of the LINE command? _______

2. What do you respond to "Specify next point or [Undo]:" prompt of the LINE command if you want to close a polygon?______

3. To zoom to the previous view, which ZOOM option would you use?______

4. What is the difference between the "window" and "window crossing" methods of selecting objects? How do you execute each method?

5. What is the difference between the OOPS and U commands? What is the different between the U and UNDO commands?

PRACTICE

1. Draw a few circles using the "2 point" and "TTR" options and enter the command by using the pulldown menus.

2. Use ORTHO and SNAP to draw a box 3' wide and 2' long.

3. Draw a box 1'3" wide and 1'8" long by entering the polar coordinates.

ADDITIONAL WORK

More ZOOM Command Options

Another ZOOM option is just to specify a number for a scale:

> Command: **Z**
> ZOOM
> Specify corner of window, enter a scale factor (nX or nXP), or
> [All/Center/Dynamic/Extents/Previous/Scale/Window] <real time>:**1.5**

You should be zoomed in closer. Specifying less than 1 will bring the drawing farther out. Try different numbers below:

1. **Use the ZOOM command with the numbers: 2, 3, 4.**
2. **ZOOM out to drawing limits .**
3. **ZOOM with .75 to see drawing from farther out.**
4. **ZOOM out to drawing limits . .**

Another ZOOM option is to specify a center point then specify a height for the zoom window:

Command: **(press ENTER or right click)**
ZOOM
Specify corner of window, enter a scale factor (nX or nXP), or
[All/Center/Dynamic/Extents/Previous/Scale/Window] <real time>: **C**
Specify center point: **(point in upper right quadrant of drawing)**
Enter magnification or Height (11'-4"): **4' (or other height)**

Note that the 4' height window is centered around the the point you selected.

BLIPMODE Command

Blips are drawing markers. They appear as little crosses where you have selected points. Do not turn them on for this tutorial but if sometime you do want to turn on to see where you have drawan, do the following:

Command: **BLIPMODE**
Enter mode [ON/OFF] <OFF> ON

TUTORIAL 3

WORKING WITH TEXT

Commands Learned:	TEXT	MTEXT
	DTEXT	DDMODIFY
	DDEDIT	QTEXT
	STYLE	

This second drawing will give you experience with the text commands. In addition you will learn how to modify properties of the text. In this tutorial, you can place the text any place within the drawing limits.

1. Click on the AutoCAD icon to load AutoCAD. If you are already in AutoCAD, select "New" under "File" on the menu bar.
2. On the Start up menu, click on "Start from Scratch" and then "OK."

Set up the drawing as below (refer back to the previous tutorial if you do not remember):

UNITS	**Architectural**
LIMITS	**upper right corner 44',34'**
GRID	**1'**
ZOOM	**all**

1. SAVE your drawing with your intitials and tut3 such as "jmktut3."

TEXT, DTEXT Commands

Both the TEXT and DTEXT commands allow you to enter text in any size and in many ways. The TEXT command will show the text only after you have finished the command while DTEXT lets you see the text while you are typing it, hence it is referred to as dynamic text. You will use DTEXT most of the time so the instruction below uses that command. Use the "Caps Lock" key to make the text all capital letters. First, you will use the menu bar to enter text:

1. Menu Bar: Draw > Text > Single Line Text.

The response will be:

DTEXT

Current text style: "STANDARD" Text height: 0’-0 3/16”
Specify start point of text or [Justify/Style]: **(point to where you want text to start)**
Specify height <1'-0">: **1'**
Specify rotation angle of text <0>: **(press ENTER)**
Enter text: **Entering TEXT is easy!**
Enter text: **(press ENTER)**

The text will be entered left justified. You specified 1' so that when the paper is printed out at 1/4"=1’ scale the text will actually be 1/4” in height. If you had put 1/4" the text would be almost invisible on the screen; however, if you zoom in very close you could find it to erase. Next, you will use the ENTER key to repeat the command and to place text under the previous text:

Command: **(press ENTER or right click)**
DTEXT
Current text style: "STANDARD" Text height: 1'-0"
Specify start point of text or [Justify/Style]: **(press ENTER)**
Enter text: **CAD**
Enter text: **(press ENTER)**

To change text height and angle, you will need to type DTEXT to be prompted for these options:

Command: **DTEXT**
Current text style: "STANDARD" Text height: 1'-0"
Specify start point of text or [Justify/Style]: **(point anywhere graphics window)**
Specify height <1'-0">: **2'**
Specify rotation angle of text <0>: **90**
Enter text: **Designer**
Enter text: **(press ENTER)**

You must change the rotation angle back to 0 to make text horizontally.

1. **Enter some more text in various sizes and rotations (Fig. 3-1).**

Fig. 3-1 Entering text with DTEXT varying the height and rotation.

You can “right justify” the text by entering the R option:

Command: **(press ENTER or right click)**
DTEXT
Current text style: "STANDARD" Text height: 2'-0"
Specify start point of text or [Justify/Style]: **R**
Specify right endpoint of text baseline: **(point anywhere in graphics window)**
Specify height <2'-0">: **1'**
Specify rotation angle of text <90>: 0
Enter text: **TEXT IS JUSTIFIED RIGHT**
Enter text: **(press ENTER)**

The DTEXT commands have a number of justify options which can be seen by entering “J” to the first prompt. In the next sequence, you will center text and then enter muliple lines of centered text. However, AutoCAD will not center the text until after you end the command (Fig. 3-2).

Command: **DTEXT**
Current text style: "STANDARD" Text height: 1'-0"
Specify start point of text or [Justify/Style]**: J**
Enter an option [Align/Fit/Center/Middle/Right/TL/TC/TR/ML/MC/MR/BL/BC/BR]: **C**
Specify center point of text: **(point anywhere in graphics window)**
Specify height <1'-0">: **(press ENTER)**
Specify rotation angle of text <0>: **(press ENTER)**
Enter text: **DESIGNERS INC.**
Enter text: **2405 West Road**
Enter text: **Hometown, USA**
Enter text: **(press ENTER)**

Fig. 3-2 Centering text with DTEXT command.

DDEDIT Command

The DDEDIT command allows you to edit a line of text similar to a word processor in a dialogue box.

1. Menu Bar: Modify > Text.

The response will be the same as the second line below.

Command: **DDEDIT**
<Select a TEXT or ATTDEF object>/Undo: **(point to a line of text)**

A dialogue box appears. You can use the backspace or delete key to delete the text or use the cursor keys to back up to where you want to delete or replace text.

STYLE Command

With the STYLE command you can define many different styles. In the following illustration are some typical fonts (Fig. 3-3). Two fonts, Monotext and Romans, are simple AutoCAD fonts that are drawn quickly. Romans is useful for entering text on architectural drawings because it can be read even when it is very small. Monotext is the only font that does not use proportional spacing--useful when you want to line up text in schedules. The other fonts shown are TrueType font equivalents. These are most useful for presentation type drawings or in title blocks.

ARIAL	ABCDEFGHIJKLMNOPQRSTUVWXYZ123456789
ASIANA	**ABCDEFGHIJKLMNOPQRSTUVWXYZ123456789**
BRECTANGULAR	**ABCDEFGHIJKLMNOPQRSTUVWXYZ123456789**
CITYBLUEPRINT	ABCDEFGHIJKLMNOPQRSTUVWXYZ123456789
COUNTRYBP	ABCDEFGHIJKLMNOPQRSTUVWXYZ123456789
DUTCH801 RM BT	*ABCDEFGHIJKLMNOPQRSTUVWXYZ123456789*
FUTURA MD BT	**ABCDEFGHIJKLMNOPQRSTUVWXYZ123456789**
GALLIARD BT	*ABCDEFGHIJKLMNOPQRSTUVWXYZ123456789*
HIGHLANDER	**ABCDEFGHIJKLMNOPQRSTUVWXYZ123456789**
MISTRAL	*ABCDEFGHIJKLMNOPQRSTUVWXYZ123456789*
MONOTEXT	ABCDEFGHIJKLMNOPQRSTUVWXYZ123456789
PLEASANT	*ABCDEFGHIJKLMNOPQRSTUVWXYZ123456789*
ROMANS	ABCDEFGHIJKLMNOPQRSTUVWXYZ123456789
SWIS721	**ABCDEFGHIJKLMNOPQRSTUVWXYZ123456789**
TECHNICAL	ABCDEFGHIJKLMNOPQRSTUVWXYZ123456789
TIMES	ABCDEFGHIJKLMNOPQRSTUVWXYZ123456789

Fig. 3-3 Some typical text styles.

AutoCAD has the standard font set up for the text style. To use the other fonts, you must first name the style and then assign a font file to that name. Then, all text entered after this procedure will be in this font which is the current font style until you change it again. You will do this in the following procedure.

1. Menu Bar: Format > Text Style..."
2. In the Text Style window click on "New..."
4. Type in "romans"
5. Under font name, scroll up and highlight "romans.shx" to load it.

You are able to see the style in the "Preview" section. The "Text Style" dialogue window allows you to enter a height, but if you do that you will not be prompted for a height when using the DTEXT command. Unless you know you will only use the same height with a style, leave height at 0. Likewise, you will probably not want to change any of the effects such as backwards, upside down, etc. unless you want to experiment with changing the font geometry. You will now exit this dialogue window:

6. Click on "Apply" button.
7. Click on "Close" to close the dialogue window.

Then enter some more text to see this text style:

1. Menu Bar: Draw > Text > Single Line Text.
2. Enter some text as before.
3. Using the above procedure, make another text style.
4. Enter more text.

SPELL Command

The SPELL command checks spelling.

1. Use DTEXT to enter a line of text with at least one misspelled word.
2. Menu Bar: Spelling
3. Highlight correct word and click on "Change."

MTEXT Command

The MTEXT command allows you to enter text in paragraph form. You are first prompted to make a window or box that will enclose the text. A dialogue window appears that allows you to change the style of all the text, import text files, or change style or font of individual words. Experimenting with this dialogue window is the best way to learn this useful feature. MTEXT is easily used by clicking on the icon on the draw toolbar. The two corners create the space that will contain the text.

Command: **MTEXT**
Current Text Style: ROMANS Text Height: 1'
Specify first corner: **(point in lower corner)**
Specify opposite corner or [Height/JustifyLine spacing//Rotation/Style/Width]: **(point in upper right)**

The dialogue box appears (Fig. 3-4).

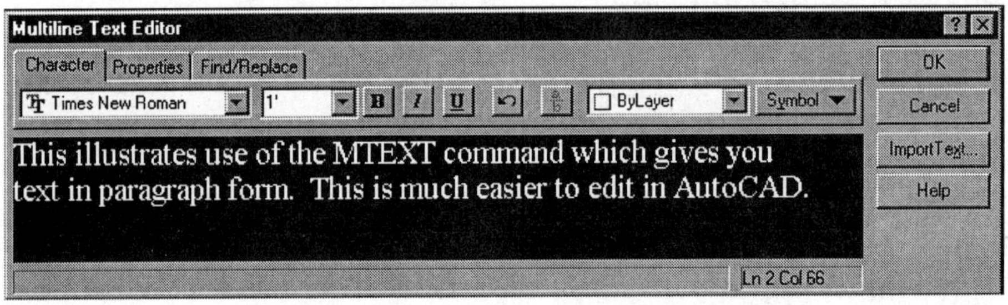

Fig. 3-4 MTEXT dialogue box.

1. **Enter text of at least 15 words.**
2. **Click on "Properties" to bring up another window.**
3. **Change the text style.**
4. **Click back on "Character" window.**
5. **Select a word by highlighting and clicking the U button to underline.**
6. **Highlight another word and change the color or text style.**

The bold button only works on selected truetype fonts (those with the double t icon next to them). MTEXT has several advantages over the DTEXT command. You can import text from a "txt" file which is easily made with a word processing program. Cut-Paste is also easily done in the Multiline Text Editor. You can change text styles without having to set a new text style. Using the DDEDIT command on text entered with the MTEXT command brings up the Multiline Text Editor allowing you not only to change text but other properties. MTEXT window has more advanced features such as the "Find/Replace" Tab which allows you to find words and replace with other words.

Another advantage of the MTEXT command is that you can change the shape of the window of the text or the two points you entered orignally. You do this by using grips. You will learn more about using grips later. You will try this feature by moving the text. You first will make sure that the grips feature is turned on or GRIPS is equal to 1:

Command: **GRIPS**
Enter new value for GRIPS<0>: **1**

1. **Make a window around the text—you should see 4 blue boxes.**
2. **Click on one corner to highlight the box in red.**
3. **Then move the crosshair to a new location.**

DDMODIFY Command

The DDMODIFY command allows you to change properties of any object:

1. **Menu Bar: Modify > Properties.**

The Properties window appears with “No Selection” appearing at the top.

2. **Click on a line of text entered with DTEXT.**
3. **Click on Height under Text in Properties window and change height.**

This important command will be introduced again in a later tutorial.

SHORT ANSWER/DISCUSSION QUESTIONS

1. What command do you use to enter paragraph text?____________________

2. How do you establish a new text style?

3. Suppose you had entered text with the Roman Simplex style and wanted to change it to a new style, what command would you use to do this?

____

4. What command should you use to edit a line of text?__________

PRACTICE

1. Practice with the text and text editing.

2. Make a poster or design with text (such as a favorite quote) and two-dimensional design with commands learned in the previous exercise.

ADDITIONAL WORK

More Justification Options for Text

For the align, fit, center, middle, and right justification, AutoCAD places the text right above the point you select. The TL, ML, and BL options will place text left justified but at the top, middle, and bottom of the text. The following illustration (Fig. 3-5) shows text entered in relation to the points when you use the BC (Bottom Center), MC (Middle Center), or TC (Top Center Options).

Fig. 3-5 You can precisely place letters with the TL, ML, and BL options.

Aligning the text along a line or entity may be useful. AutoCAD will vary the height depending on how much text is typed and the distance between the two endpoints (Fig. 3-6):

1. Use the LINE command to draw a diagonal line.

Command: **DTEXT**
Current text style: "STANDARD" Text height: 1'
Specify start point of text or [Justify/Style]: **A**
Specify first endpoint of text baseline: **(point a little above one end of line)**
Specify second endpoint of text baseline: **(point a little above other end)**
Enter text: **AUTOCAD**
Enter text: **(press ENTER)**

Fig. 3-6 Aligning text along a line.

The "Fit" option is similar (Fig. 3-7). This option is useful to stretch text. The sequence is:

Command: **DTEXT**
Current text style: "STANDARD" Text height: 1'-0"
Specify start point of text or [Justify/Style]: **F**
Specify first endpoint of text baseline: **(point A)**
Specify second endpoint of text baseline:**(point B)**
Specify height <1'-0">: **(press ENTER)**
Enter text: **AUTOCAD**
Enter text: **(press ENTER)**

1. Repeat the DTEXT command but when DTEXT prompts with "Text:" try pointing to a new starting point.

Fig. 3-7 Fitting text between two points.

QTEXT Command

With today's fast computers, this command might not be needed. But if your drawing contains a great deal of text (as in schedules, legends, etc.) in a complex style, it will take too long to regenerate the text in a drawing. You can use the QTEXT command to have AutoCAD put boxes where the text is. The sequence is:

Command: **QTEXT**
Enter mode [ON/OFF] <OFF>: **On**
Command: **REGEN**

If you were to print out, you would also get the boxes so make sure you turn the text back on by turing QTEXT off:

Command: **QTEXT**
Enter mode [ON/OFF] <OFF>: **OFF**
Command: **REGEN**

TUTORIAL 4

ADVANCED EDIT AND DRAW COMMANDS

Commands Learned:	ARRAY	LENGTHEN	SPLINEDIT
	BREAK	PLINE	OSNAP
	TRIM	PEDIT	
	EXTEND	SPLINE	

This tutorial concentrates on learning more edit and draw commands including learning the important object snap feature.

1. Menu Bar: File > New.
2. Click on "Start from Scratch."
3. Save the drawing as with your initials and tut4 for the filename.

Set up the drawing for 1/4"=1' scale and A size, 8-1/2" x 11" paper:

UNITS	**Architectural**
LIMITS	**upper right corner 44',34'**
GRID	**1'**
ZOOM	**all**

ARRAY Command

The ARRAY command is useful to draw interesting two-dimensional designs or to enter arrays of building elements (e.g. ceiling grids). You will first draw a small design first:

1. Zoom into an area in the lower left of the drawing limits .

2. Use the LINE command to make a design 1' square (Fig. 4-1).

3. ZOOM all to see the drawing limits again.

Fig. 4-1 Small design about 1' square.

Then make a rectangular array of this design as below. Click on the array icon on the draw toolbar or type at the command prompt:

Command: **ARRAY**
Select objects: **all**
Select objects: **(press ENTER)**
Enter the type of array [Rectangular/Polar] <R>: **R**
Enter the number of rows (---) <1>: **12**
Enter the number of columns (|||) <1>: **8**
Enter the distance between rows or specify unit cell (---): **9**
Specify the distance between columns (|||): **1'**

Fig. 4-2 Design made with the ARRAY command.

Your array may appear different than illustrated (Fig. 4-2). Notice that the rows are the horizontal direction while the columns are the vertical direction. The copy starts from the bottom up and is drawn to the right. Thus, the object you selected will be in the lower left corner. If you wanted to draw from the top down, put a negative for the unit distance between the rows (i.e. -1'). If you want to draw to the left, put a negative for the unit distance between the columns (i.e. -2').

A circular array is also useful for many applications. For the following, make up your own design.

1. Using any draw commands, make another design to occupy a little over 2' square area and in the approximate location as shown (Fig. 4-3).

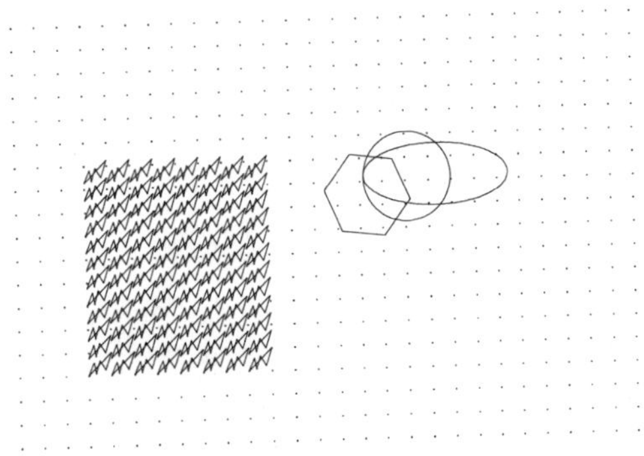

Fig. 4-3 Draw a small design to use with a circular array.

Command: **ARRAY**
Select objects: **(select all entities in your design)**
Select objects: **(press ENTER)**
Enter the type of array [Rectangular/Polar] <R>: **P**
Enter the number of rows (---) <1>: **12**
Enter the number of columns (|||) <1>: **8**
Enter the distance between rows or specify unit cell (---): **9**
Specify the distance between columns (|||): **1'**

Your design may appear similar to the one illustrated (Fig. 4-4). An angle less than 360 would have less than a full circle of objects. If you select a center point of the array too far from the object, the objects will be arrayed off the screen.If this happened, use the U command to undo the array.

Fig. 4-4 Circular design with ARRAY command.

BREAK Command

The BREAK command is important to “break out” lines, arcs, or other objects:

1. Draw a line with the LINE command.

2. Click on the break icon  **on the modify toolbar or enter:**

Command: **BREAK**
Select object: **(point to first point you want to break or A)**
Enter second break point or (First point): **(point B)**

Fig. 4-5 BREAK can be used to make an opening in a line.

TRIM Command

The TRIM command is useful for trimming out walls and other lines.

1. Draw lines as illustrated (Fig. 4-6).

2. Click on the trim icon  **on the modify toolbar or enter:**

Command: **TRIM**
Select cutting edge(s) . . .
Select object: **(point to line at A)**
Select object: **(press ENTER or right click)**
Select object to trim or [Project/Edge/Undo]: **(point to line at B)**
Select object to trim or [Project/Edge/Undo]: **(point to line at C)**
Select object to trimor [Project/Edge/Undo]: **(press ENTER or right click)**

1. Redraw the screen with the  **icon.**

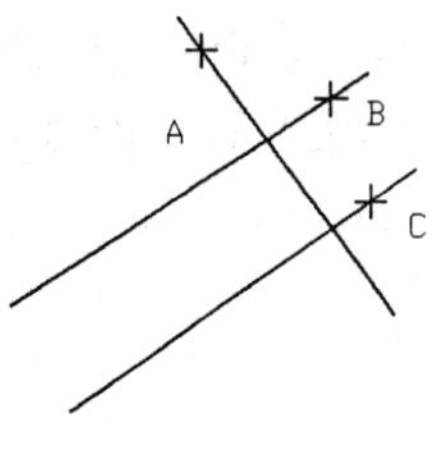

Fig. 4-6 Points for trimming.

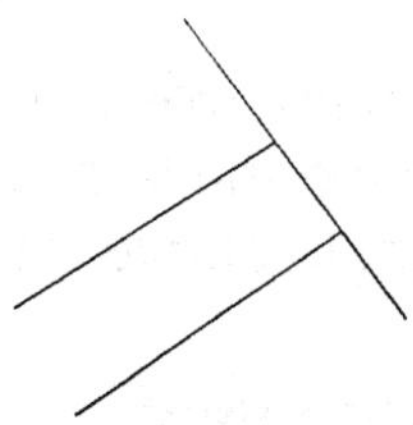

Fig. 4-7 After trimming.

EXTEND Command

This command is used to extend a line to a boundary edge. Note that the entity must be able to intersect the boundary for this command to work.

> **1. Draw two lines with the LINE command and an arc with the ARC command similar to the illustration (Fig. 4-8).**

Then, use the extend icon or type below to extend the line and arc to the horizontal line (fig. 4-9):

Command: **EXTEND**
Select boundary edge(s) . . .
Select object: **(point to horizontal line)**
Select object: **(press ENTER or right click)**
Select object to extend or [Project/Edge/Undo]: **(point to vertical line)**
Select object to extend or [Project/Edge/Undo]: **(point to one end of arc)**
Select object to extend or [Project/Edge/Undo]: **(point to other end of arc)**
Select object to extend or [Project/Edge/Undo]: **(press ENTER or right click)**

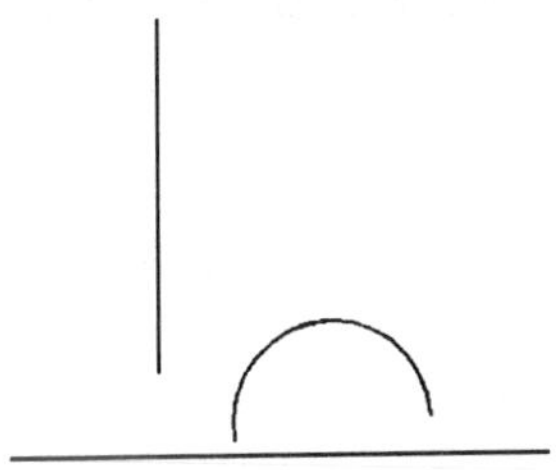

Fig. 4-8 Arc to extend.

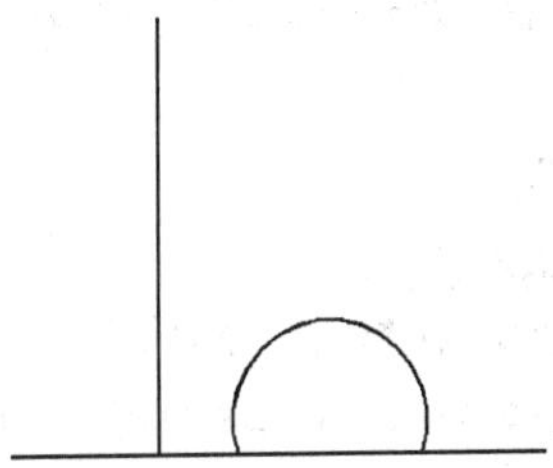

Fig. 4-9 Extended arc.

LENGTHEN Command

The LENGTHEN command allows you to lengthen a line or object. You may want to experiment with this command later but for now, just try the dynamic option. The lengthen icon is found right above the trim icon on the modify toolbar.

> **1. Draw a line anywhere.**

Command: **LENGTHEN**
Select an object or [DElta/Percent/Total/DYnamic]: **DY**

Select object to change or [Undo]: **(point on line and drag to new end point)**
Specify new end point. **(point on line and drag to new end point)**
Select an object to change or [Undo]: **(press ENTER)**

PLINE Command

The PLINE command will draw lines which have special properties and this command can be used to draw unusual shapes and lines. Polylines are treated as one entity no matter how many line segments are executed with one use of the PLINE command. The best way to understand their use is to experiment with the command and its options. In the next sequence, just point anywhere on the screen. Enter at the command prompt or use the icon on the draw toolbar:

1. Use the ZOOM command to zoom in to a small area of the drawing.

Command: **PLINE**
Specify start point: **(point)**
Current line-width is 0'-0"
Specify next point or [Arc/Close/Halfwidth/Length/Undo/Width]: **(point)**
Specify next point or [Arc/Close/Halfwidth/Length/Undo/Width]: **(point)**
Specify next point or [Arc/Close/Halfwidth/Length/Undo/Width]: **(point)**
Specify next point or [Arc/Close/Halfwidth/Length/Undo/Width]: **(point)**
Specify next point or [Arc/Close/Halfwidth/Length/Undo/Width]: **(press ENTER)**

1. Use the MOVE command and select by pointing on one polyline.

Notice that all the line segments are selected. You will change 7the starting and ending width of the line:

Command: **PLINE**
Specify start point:From point: **(point)**
Current line-width is 0'-0"
Specify next point or [Arc/Close/Halfwidth/Length/Undo/Width]: **W**
Specify starting width <0'-0">: **1**
Specify ending width <0'-1">: **12**
Specify next point or [Arc/Close/Halfwidth/Length/Undo/Width]: **(point)**
Specify next point or [Arc/Close/Halfwidth/Length/Undo/Width]: **(point)**
Specify next point or [Arc/Close/Halfwidth/Length/Undo/Width]: **(press ENTER)**

If you keep pointing, the next segments will be one foot wide. The Undo option works the same as it did with the LINE command. Likewise, the Close works the same as the LINE command when you close a polygon. End the straight lines and try drawing some thick arcs:

Command: **PLINE**
Specify start point:From point: **(point)**
Current line-width is 1'-0":
Specify next point or [Arc/Close/Halfwidth/Length/Undo/Width]: **W**
Specify starting width <0'-0">: **1**
Specify ending width <0'-1">: **1**
Specify next point or [Arc/Close/Halfwidth/Length/Undo/Width]: **A**
Specify next point or [Arc/Close/Halfwidth/Length/Undo/Width]: **(point)**
Specify next point or [Arc/Close/Halfwidth/Length/Undo/Width]: **(press ENTER)**

Once you invoke the arc option of the PLINE command, the arc mode allows you several options. Listed below are these options but you need not try them:

Angle: specify the included angle.
CEnter: specify the center of the arc.
CLose: polyline will be closed with an arc.
Direction: rather than drawing the arc tangent to the last arc segment, the direction option lets you establish the direction by pointing.
Line: switch back to line mode.
Radius: specifies the radius of the arc.
Second pt: specifies the second and third points of the arc identical to the three-point arc you have drawn previously.

As with the LINE command, you can use SNAP or ORTHO to draw orthogonally. In the illustration below, the triangle was drawn by entering 0 for starting width and 12 for ending width. The "horn" was done the same using the Arc option.

1. Practice to get shapes similar to the illustration (Fig. 4-10).

Fig. 4-10 Shapes made with PLINE command.

PEDIT Command

Because polylines are entities with complex properties, they have their own edit command called PEDIT. Do the following work so that you can use this command:

1. Use the PLINE command, set the width to 2 for starting and ending width, and draw one polyline as illustrated (Fig. 4-11).
2. Use the COPY command to make a copy of this polyline as illustrated (Fig. 4-11).

Fig. 4-11 Draw a line and copy it.

If you find you use it a lot, you will want to use the pedit icon but it is not on the modify toolbar. You will learn how to display another toolbar "Modify2" to display this icon.

1. Menu Bar: View > Toolbar...
2. Scroll down and click on "Modify II" box.
3. Click on "Close" to close the dialogue window.

You can move the toolbar around to where it is not in the way. Notice that there is the ddedit icon  that you could have used in the tutorial 3 to edit a line of text.

The top polyline will be edited with the Fit curve option which will form a smooth curve through each vertex. The bottom polyline will be edited with the Spline curve which results in a much smoother curve. Use the pedit icon or type:

Command: **PEDIT**
Select polyline: **(point to top pline)**
Enter an option [Close/Join/Width/Edit vertex/Fit/Spline/Decurve/Ltype gen/ Undo]: **F**
Enter an option [Close/Join/Width/Edit vertex/Fit/Spline/Decurve/Ltype gen/ Undo]: **(press ENTER or right click)**
Command: **(press ENTER or right click)**

PEDIT Select polyline: **(point to bottom polyline)**
Enter an option [Close/Join/Width/Edit vertex/Fit/Spline/Decurve/Ltype gen/Undo]: **S**
Enter an option [Close/Join/Width/Edit vertex/Fit/Spline/Decurve/Ltype gen/Undo]: **(press ENTER or right click)**

The result should be similar to the illustration (Fig. 4-12).

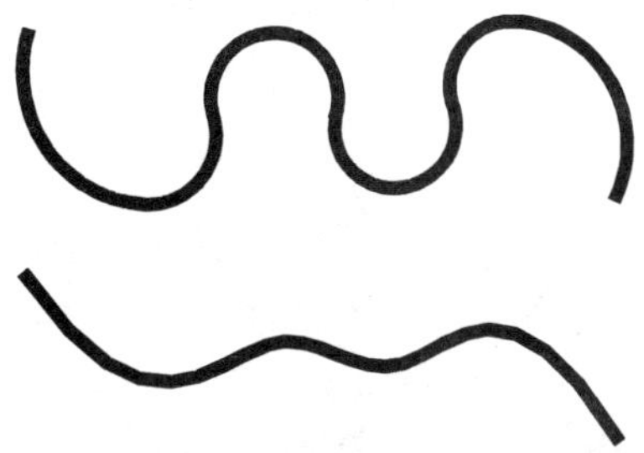

Fig. 4-12 Edited polylines.

Because you have not exited the PEDIT command, both the Undo and Decurve option would undo the curve on the bottom polyline. The other options are explained below:

Close: connects the first segment with the last segment.
Open: removes the closing segment of the polyline.
Join: will add separate polylines as well as lines and arcs and add them to a polyline to be treated as one polyline or entity.
Width: specifies a new uniform width for the selected polyline.
Edit vertex: with a selected vertex, you can edit in a variety of ways including moving, breaking, and straightening.

Other Uses for PEDIT Command

The PEDIT command can be used to convert arcs or lines to polylines. This might be useful when you can draw something with the ARC or LINE command but you want the shape to have a thickness. You would first use the PEDIT command to convert the arc to a polyline, then join the other lines to the arc, then change the thickness.

SPLINE Command

The “Fit curve” and “Spline curve” options of PEDIT are useful to make curved lines with thickness. However, if you just want a single thickness curved line, the SPLINE command allows you to draw these directly. Try to draw a similar shape as in the figure (Fig. 4-13). Use the spline icon on the draw toolbar or enter:

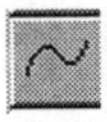

Command: **SPLINE**
Specify first point or [Object]: **(point)**
Specify next point: **(point)**
Specify next point or [Close/Fit tolerance] <start tangent>: **(point)**
Specify next point or [Close/Fit tolerance] <start tangent>: **(point)**
Specify next point or [Close/Fit tolerance] <start tangent>: **(point)**
Specify next point or [Close/Fit tolerance] <start tangent>: **(point)**
Specify next point or [Close/Fit tolerance] <start tangent>: **(point)**
Specify start tangent: **(point)**
Specify end tangent: **(point)**

Fig. 4-13 Curve made with SPLINE command.

SPLINEDIT Command

Similar to the PEDIT command, splines have their own edit command—SPLINEDIT. You will use this command to move the first vertex on the spline. But first you will use “Fit data” option to fit the control points or vertices on the spline. Your edited spline may look similar to illustration (Fig. 4-14). Use the menu bar or type:

1. Menu Bar: Modify > Spline.

Command: **SPLINEDIT**
Select spline: **(point on spline)**
Enter an option [Fit data/Close/Move vertex/Refine/rEverse/Undo]: **F**
Enter a fit data option

[Add/Close/Delete/Move/Purge/Tangents/toLerance/eXit] <eXit>: **M**
Specify new location or [Next/Previous/Select point/eXit] <N>: **(point to new location of vertex)**
Specify new location or [Next/Previous/Select point/eXit] <N>: **X**
Enter a fit data option
[Add/Close/Delete/Move/Purge/Tangents/toLerance/eXit] <eXit>: **(press ENTER or right click)**
Enter an option [Fit data/Close/Move vertex/Refine/rEverse/Undo]: **(press ENTER or right click)**

Fig. 4-14 Edited spline with SPLINEDIT command.

Using GRIPS With Polylines

Grips, introduced previously, can be used as an intuitive way to edit objects and increase your productivity. You will use grips to edit a vertex.

1. **Draw a polyline with vertices similar to the illustration (Fig. 4-15).**
2. **Make a window around the polyline.**

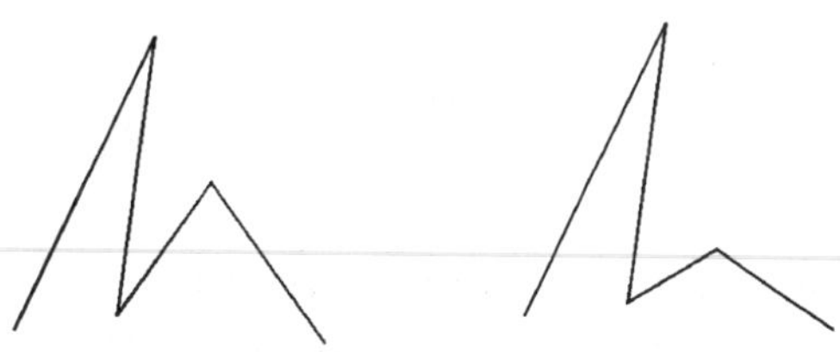

Fig. 4-15 Polyline before and after use of grips to move vertex.

You should see the polyline appear dashed because it has been selected and see blue edit boxes or grips on each vertex.

1. **Click on a vertex until it is highlighted or turns red.**
2. **Then move the vertex to a new location by moving the crosshairs.**

If you looked at the command prompt while you were doing this, you would have seen that you were using the STRETCH command to move the vertex to a new location. We will learn about the STRETCH command in a later exercise.

1. Repeat the above sequence but this time, press ENTER when AutoCAD prompts with the STRETCH prompt.
2. When MOVE comes up, point to a new location.

The third option is to rotate the polyline using the selected vertex as the axis of rotation. The grips feature can be used to edit more objects. If you make the window from upper right to lower left corner, it is the same as window crossing that you learned before.

1. Draw a circle with the CIRCLE command or use 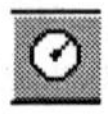 icon.

The grips are located at the center and on the four quandrants of the circle.

2. Click on the center grip to move the circle.
3. Click on a quadrant grip and move it to make the diameter larger.
4. Draw an arc with the ARC command.
5. Use grips to change the endpoint of the arc.

You can disable the grips feature by changing the value of the system variable GRIPS to 0; however, do not do this unless they really bother you.

```
Command: GRIPS
Enter new value for GRIPS <1>: 0
```

1. SAVE this drawing file.

You will start a new drawing for the rest of this tutorial.

Using Object Snap to Draw a Rug Design

The objective of the rest of this tutorial is to learn the use of object snap and hatch. Because AutoCAD is based on vectors and coordinate geometry, it can easily find coordinate pairs. You tell AutoCAD the location with use of object snap. Listed below are the different types of object snap and description of what AutoCAD will do:

endp	Endpoint of a line
mid	Midpoint of line
int	Intersection of two lines
app	Apparent intersection (used most in 3D)
cen	Center point of circle or polygon
quad	Quadrant of circle (4 equidistant points on circle circumference)
per	Next point will be drawn perpendicular to object
tan	Tangent to object
node	Point location (introduced in later exercise)
ins	Insertion point of a block (introduced in later exercise)
nea	Nearest point

1. **On the menu bar, click on "File", "New" and "Start from Scratch."**
2. **Enter file name with your initials and "rug" such as "jmkrug" in the drawing file box.**

You will use most of these in making your rug design (Fig. 4-16) and the tutor will take you through it step by step.

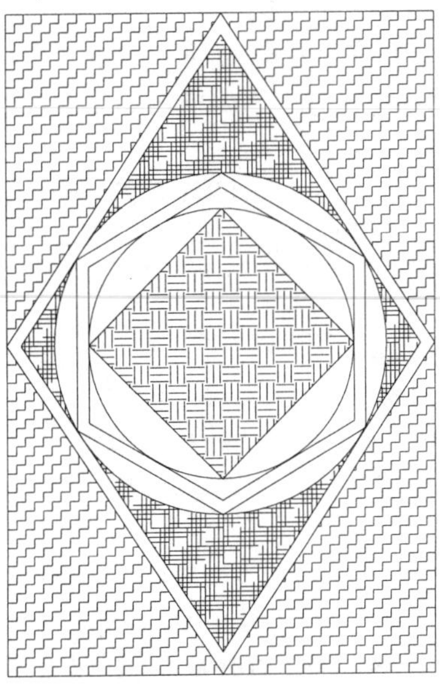

Fig. 4-16 Rug design facilitated by use of object snap.

You will set up the drawing for 1"=1'scale and letter-size or A-size paper.

UNITS	**Architectural.**
LIMITS	**upper right corner at 8',11'**
GRID	**1'**
ZOOM	**all**

You will turn the UCS icon off and draw the outline of the 6' x 9' rug by entering polar coordinates:

Command: **UCSICON**
Enter an option [ON/OFF/All/Noorigin/ORigin] <ON>: **OFF**
Command: **L**
LINE Specify first point: **1',1'**
Specify next point or [Undo]: **@6'<0**
Specify next point or [Undo] **@9'<90**
Specify next point or [Close/Undo] **@6'<180**
Specify next point or[Close/Undo] **C**

OSNAP Command

The OSNAP command establishes a "running" object snap mode, letting you repeatedly snap to what you select (i.e. endpoint, intersection, center, etc). You can also choose multiple object snap setting. OSNAP is found under the menu bar "Tools" but you also can use the object snap toolbar.

1. On the menu bar, click on "View"
2. Click on "Toolbars..."
3. Scoll down and click on the box next to "Object Snap" until an x appears.

Command: OSNAP

1. In the dialogue box, click on the box next to "Midpoint" and "Intersection" until an x appears in each box.
2. Click on "OK" to exit the dialogue window.

OSNAP can be turned off by clicking on OSNAP on the status line. Until you change this mode, you will only be allowed to draw from the midpoint of a line or the intersections of two lines. AutoCAD will show a midpoint cursor icon (triangle) or intersection (cross) to show that you are snapping to either of these objects. Next, you will use this object snap mode to draw a line to the midpoint of each line resulting in a diamond shape (Fig. 4-17):

Command: **PLINE**
Specify start point: **(point A)**
Current line-width is 0'-0"
Specify next point or [Arc/Close/Halfwidth/Length/Undo/Width/]: **(point B)**
Specify next point or [Arc/Close/Halfwidth/Length/Undo/Width/]: **(point C)**
Specify next point or [Arc/Close/Halfwidth/Length/Undo/Width/]: **(point D)**
Specify next point or [Arc/Close/Halfwidth/Length/Undo/Width/]: **C**

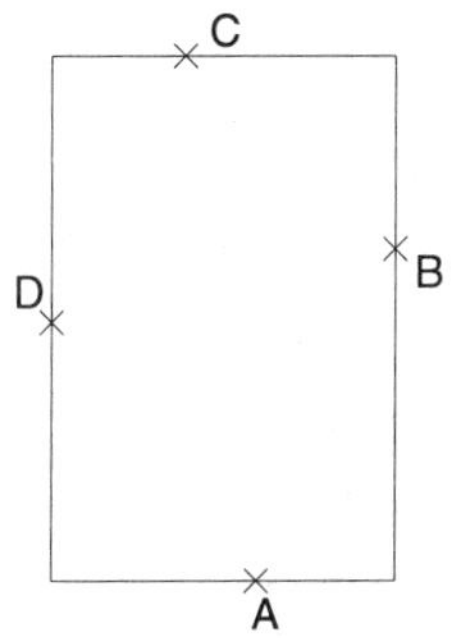

Fig. 4-17 Click anywhere along each line for midpoints.

OFFSET Command

This command will offset a copy of an entity at a distance you specify.

Command: **OFFSET**
Specify offset distance or [Through] <Through>: **2**
Select object to offset or <exit>: **(point anywhere on polyline)**
Specify point on side to offset? **(point on inside of polyline)**
Select object to offset or <exit>: **(press ENTER or right click))**

The result should be similar to the illustration (Fig. 4-18).

Fig. 4-18 Use of the OFFSET command to offset the polyline.

Had you just used the LINE command to make the first diamond, the lines would have overlapped. Next you will draw a circle that is tangent to the diamond. But first you will draw a construction line by finding the intersection of the diamond and outline (Fig. 4-19):

Command: **L**
LINE Specify first point: **(point A)**
Specify next point or [Undo]: **(point B)**

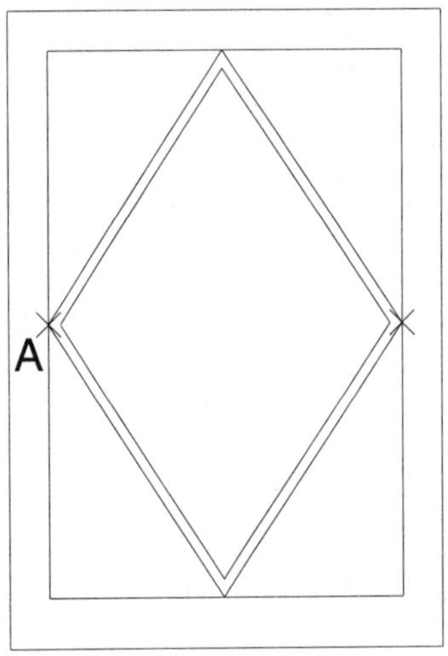

Fig. 4-19 Drawing a line from the intersection.

Next, you will clear the object snap setting:

Command: OSNAP

1. **Click on "Clear all."**
2. **Click on "OK" to exit the object snap dialogue window.**

You will click on the individual object snap icons using the midpoint and a tangent to draw the circle (Fig. 4-20):

Command: **C**
CIRCLE
Specify center point for circule or [3P/2P/Ttr/(tan tan radius): **mid (or click on** 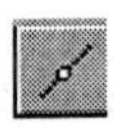 **icon)**

of **(point A)**

Specify radius [Diameter]: **tan (or click on**  **icon)**

of **(point B)**

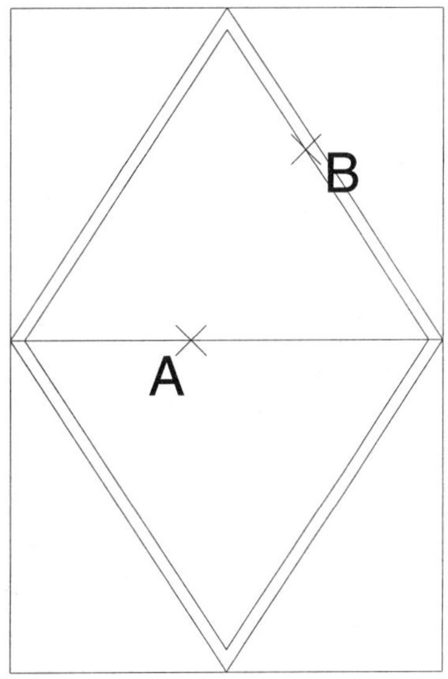

Fig. 4-20 Snapping to the midpoint.

1. Use the ERASE command to erase the construction line.

In the next sequence, you will draw a polygon inside the circle. You want it to have the same center point so you will use the "cen" object snap. You can place the edge of the polygon on the circle by using the "near" object snap option or nearest point. Finally you will draw a circle:

Command: **POLYGON**
Enter number of sides <4>: **6**
Specify center of polygon [Edge]: **cen (or click on icon)**
of **(point anywhere on circle circumference)**
Enter an option [Inscribed in circle/Circumscribed about circle]<I>: **(press ENTER)**

Radius of circle: **nea (or click on icon)**
to **(point on the bottom of the circle circumference)**

1. Use the OFFSET command to offset the polygon 2 inches inside.

Command: **C**
CIRCLE
Specify center point for circle or [3P/2P/Ttr/(tan tan radis)]: **cen (or click on icon)**
of **(point on circle-not the polygon)**
Specify radius [Diameter]: **tan (or click on icon)**
of **(point to polygon)**

Your drawing should look like the illustration (Fig. 4-21).

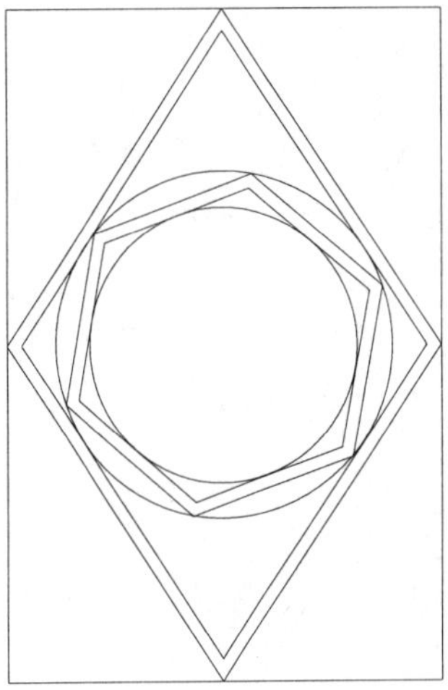

Fig. 4-21 Drawing a polygon and circle with aid of object snap.

You will set object snap to "quad" or quadrant to make a box inside the circle (Fig. 4-22).

Command: OSNAP

1. In the dialogue box, click on the box next to "Quadrant" until an x appears.
2. Click on "OK" to exit the dialogue window.

Command: **L**

LINE
Specify first point: **(point approximately at A)**
Specify next point or [Undo]: **(point approximately at B)**
Specify next point or [Undo]: **(point approximately at C)**
Specify next point or [Close/Undo]: **(point approximately at D)**
Specify next point or [Close/Undo]: **C**

Command: OSNAP

2. Click on "Clear all" and "OK" to close the dialogue box.

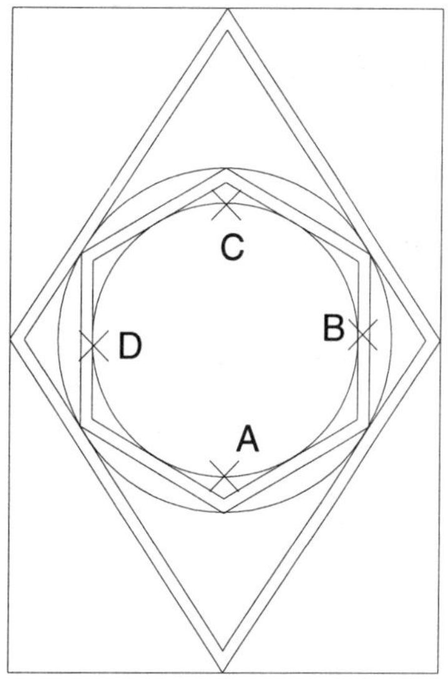

Fig. 4-22 Using quadrants of the circle to make a box inside the circle.

Next you will use hatch patterns to make your rug design more interesting and learn about the basic aspects of hatching.

HATCH, BHATCH Commands

The HATCH command draws patterns within an area or boundaries. The HATCH command requires enclosed boundary. It is easiest to use the pulldown menus to set a hatch style. Use the hatch icon on the draw toolbar or type at the command prompt:

Command: **BHATCH**

1. **Click on "Patterns".**
2. **Pull bar down until you see "earth" pattern and then click on it.**
3. **Click on the "Scale" box and enter 12 (very important!).**
4. **Click on "Select objects"**
5. **Select the inside box by selecting each line.**
6. **Press ENTER to "Select objects:" prompt.**
7. **Click on "Preview hatch."**
8. **If it looks like the illustration, click "Enter" to return to window.**
9. **Click on "OK" to finish hatch.**

The result should look like the illustration (Fig. 4-23).

Fig. 4-23 Hatching the box.

You changed the scale to 12. At the scale of 1, you would probably not see the hatching as it would be solid. Be sure to preview your hatch to make sure it is the right size. If you are unsure of the scale, specify a much larger scale. You can always scale it down later. Another aspect of scale is that it is related to your drawing limits. If you set up your drawing for 1/4"=1' scale, you should use 48 for the scale. If you set up for 1/8"=1' scale, then you should set scale at 96 and so on.

AutoCAD needs an enclosed boundary to complete a hatch. If not, AutoCAD will give up or fill it in incorrectly.

If you have an object or entity within an entity and want to hatch between the two, you can select first the outer boundary and then the internal boundary. AutoCAD will fill the hatch and stop at the boundary as below (Fig. 4-24).

1. Use the BHATCH command (or icon) to bring up the Boundary Hatch window.
2. Click on "Patterns" box.
3. Click on another pattern.
4. Check that the "Scale" box still reads 12.
5. Click on "Select objects" button.
6. Select the outer circle and the inside diamond as shown (Fig. 4-24).
7. Press ENTER to "Select Objects" prompt
8. Click on "Preview hatch" button.
9. If it looks like the illustration, click on "Enter" to return to window.
10. Click on "OK" to finish hatch.

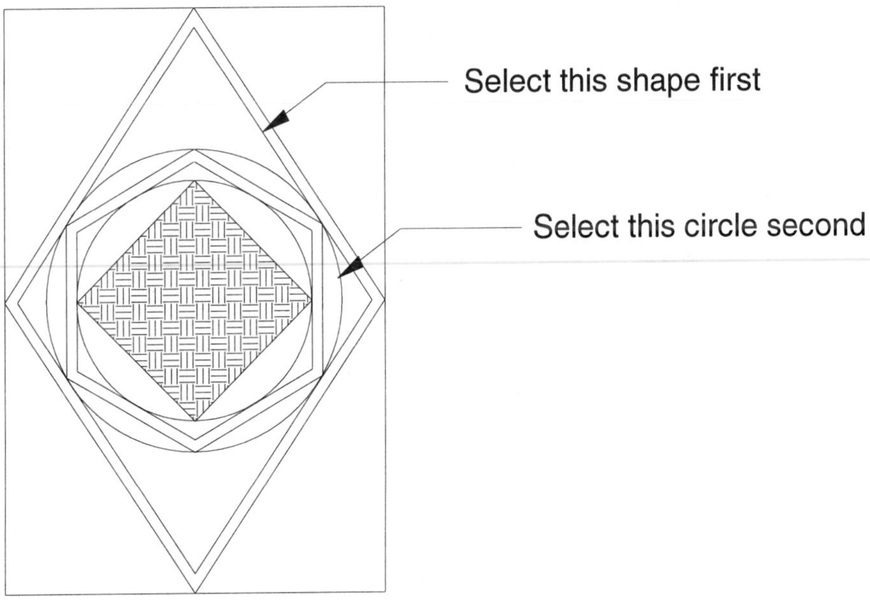

Fig. 4-24 Selecting an area between two objects.

Often you will want to label within the hatch pattern. AutoCAD will hatch around the text creating an invisible box if you select the text as a boundary. If you do not select the text, AutoCAD will ignore the text and fill in the hatch over it.

When two entities overlap, selecting the objects will not work because one object will be filled in completely. In this case, you pick points and AutoCAD will search in all directions from the point to select a boundary. Such is the case with the outer area of the rug design where the outer lines overlap the diamond (Fig. 4-25):

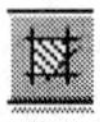

Command: **BHATCH**

1. **Click on "Patterns" to bring up the "Hatch Patterns Palette" window.**
2. **Pull bar down until you see "earth" pattern and then click on it.**
3. **Select another hatch pattern if you want.**
4. **Click on "Pick points . ." box.**
5. **Point in areas illustrated (Fig. 4-25).**
6. **Press ENTER key to get dialogue box back.**
7. **Click on "Preview Hatch" box.**
8. **After viewing, click on "Enter" to get dialogue box back.**
9. **Click on "OK" to finish hatch.**

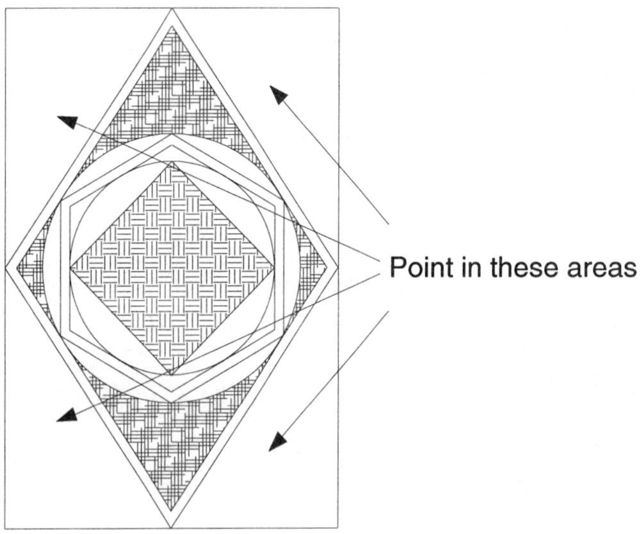

Fig. 4-25 Picking points to have AutoCAD search for boundaries for the hatch.

Plotting Your Drawing on the Printer

To print or plot your drawing, click on the following icon or enter at the command prompt:

Command: **PLOT**

This is the same as clicking "File" on the menu bar and then "Plot." A Plot dialogue window appears. The next section discusses each area within the two tabs: Plot Device and Plot Settings.

Plot Device Tab

This tab lists the current plotter configuration. If you have configured AutoCAD with more than one device, such as a plotter and a printer, you would select another plot device in the Plotter Configuration area. This is where you can specify a DWF plot file (explained later). Other parameters that you can change include plot style table to assign pens and the number of copies. The "Plot to file" box, when checked, will not plot the drawing, only write the plot to a file.

Plot Settings Tab

Paper size and paper units: Paper size is dependent on your printer or plotter while the printable area will always be less than the paper size. Units are set at inches which is most often what you want to use.

Plot area: You choose the area of the drawing to be plotted by clicking on only one button:

Limits: to print the whole paper including drawing lmits.
Extents: similar to zooming extents.
Display: what is displayed on screen gets plotted.
Window: lets you make a window around the area you want to plot.
View: allows you to call up a saved view (from the VIEW command) to plot (explained in a later tutoiral).

Drawing orientation: You can plot in portrait (vertical) or landscape (horizontal) styles.

Plot scale:

Scale: Change the scale by making a selection or entering a custom scale .25=1' gives you 1/4"=1'scale while .5=1' is 1/2"=1', etc. Alternatively you can click on the "Scaled to Fit" to have this drawing fit the page but it will not be to scale.

Plot Offset: This feature centers the plot or offset in the X or Y direction.

Plot Options: “Hide lines” is only checked when you are working in 3D.

Full and Partial Previews: this useful option allows you to preview your plot either partially (showing the area of the limits that will be plotted) or the full preview. The full preview shows what your plot will look like. Be sure to preview your plot.

NOTE: For this tutorial, you can click on “Display” and “Fit to Scale” only to plot.

Finally, click on “OK” when you have checked all parameters. This sends the plot to the printer or plotter.

Your plot settings are stored in your drawing file. Consequently, each file will have different plot settings based on the last plot you made. Later, you will learn about paper space which allows you to create different layouts with plot settings.

SHORT ANSWER/DISCUSSION QUESTIONS

1. What are some uses of the ARRAY command?

2. What command is used to edit a polyline? A spline?

3. What command could you use to make the vertical line meet the horizontal line?

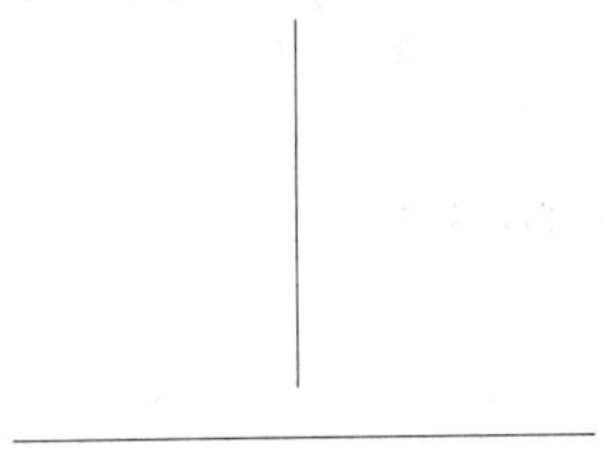

4. If you wanted to start a line the center point of a circle, what do you enter to the “Specify first point:” prompt of the LINE command?

5. Why is it critcally important to change the scale of a hatch pattern from one to a much higher number (i.e. 48) when your drawing is set up for A size and1/4”=1’ scale in model space.

PRACTICE

1. Make your own rug or textile design making use of object snap and other commands you have learned.

TUTORIAL 5

MORE DRAWING ENHANCEMENTS

Commands:	SOLID	COLOR	FILLET	MIRROR	TRACE
	DONUT	SKETCH	ROTATE	SCALE	FILL

In this tutorial, you will learn to make solid areas, sketch freehand lines, and make other irregularly shaped lines. Also, you learn a few more edit commands.

1. Start a new drawing and name it tut5 and include your initials.

Set up the drawing for 1/4" scale and letter-size paper (A-size) as below:

UNITS	**Architectural**
LIMITS	**upper right corner 44',34'**
GRID	**1'**
ZOOM	**all**

SOLID Command

The SOLID command is useful to fill in columns in a plan or whenever you need a solidly filled area. It is drawn similar to the LINE command, but you must be careful how you sequence the points (Fig. 5-1):

1. ZOOM to small area of screen (about 10' by 10' window).
2. Set SNAP to 3" and make sure it is on.

Command: **SOLID**
Specify first point: **(point A)**
Specify second point: **(point B)**
Specify third point: **(point C)**
Specify fourth point or <exit>: **(point D)**
Specify third point: **(press ENTER or right click)**

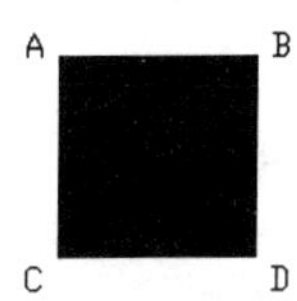

Fig. 5-1 Making a solidly filled figure with SOLID command.

Had you reversed the last two points, the solid would not have filled into a box but would appear to be a bowtie. To draw a solid triangle, enter the first two points of the edge and then one point as below (Fig. 5-2):

Command: **SOLID**
Specify first point: **(point A)**
Specify second point: **(point B)**
Specify third point: **(point C)**
Specify fourth point or <exit>: **(press ENTER)**
Specify third point: **(press ENTER)**

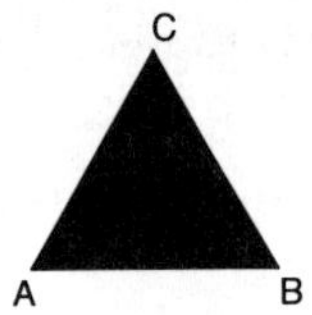

Fig. 5-2 Filled triangle.

DONUT Command

The DONUT command is useful for drawing solid circles or rings (Fig. 5-3). To make a filled circle, the inside diameter is 0.

1. On the menu bar, click on "Draw" and then "Donut" or enter at command prompt:

Command: **DONUT**
Specify inside diameter of donut <0'-0 1/2">: **0**
Specify outside diameter of donut <0'-1">: **2'**
Specify center of donut or <exit>: **(point)**
Specify center of donut or <exit>: **(point to make another)**
Specify center of donut or <exit>: **(press ENTER or right click)**

If you really wanted a donut, you would specify an inside diameter:

Command: **DONUT**
Specify inside diameter of donut <0'-0 1/2">: **2'**
Specify outside diameter of donut <0'-1">: **4'**
Specify center of donut or <exit>: **(point)**
Specify center of donut or <exit>: **(point to make another)**
Specify center of donut or <exit>: **(press ENTER or right click)**

COLOR Command

Later, you will learn about layers which is the best way to assign colors. However, once you select a color--thereafter each object will be drawn in that color until you change it.

1. Menu Bar: Format > Color...
2. Select a color and click on "OK" to exit the dialogue window.
3. Use any draw command to see that you have a different color.

Another way to change the color is to click on the Color Control on the Object Properties Toolbar. Now it should show your color--you can select quickly from the standard colors.

Filling in Other Shapes with Solid Areas---BHATCH

The BHATCH command is the easiest way to fill in a complex shape with a solid area. Likewise, you may want to just use this command to fill in circles and rectangles rather than the SOLID or DONUT command.

1. Draw some complex shapes such as a polygon.
2. Use BHATCH and select "Solid" for the pattern to hatch an area.

SKETCH Command

The SKETCH command is used to draw freehand lines. The freehand line is made up of line increments as you sketch with the pointing device. The larger the line increment, the more jagged the line will appear; conversely, the smaller the line increment, the smoother the line. However, smaller line increments will mean that the freehand line is composed of many lines which can qucikly make a large file size. Use this command only when it is absolutely necessary and then very cautiously. It is generally not recommended.

1. Turn SNAP off on the status line or use F9 key.

Command: **SKETCH**
Record increment <0'-0 1/8">: **1"**
Sketch. Pen eXit Quit Record Erase Connect . **(hold mouse button down and move it)**

After you point with the pointing device, the pen is considered to be down and you can draw similar to using a pencil. To raise the pen up or stop sketching, point again. To start again, point. To exit, enter "X" which will return you to the command prompt. AutoCAD records and redraws the lines and gives a message similar to below:

164 lines recorded

The line you just drew is made up of 164 individual lines which can be edited individually. If you wanted to erase these, the best way would be to put a window around all of the lines or use the UNDO command.

An explanation of the other subcommands to the SKETCH command appears below for reference only.

Quit: If you want to quit while you are sketching, type Q for quit or press Esc key.
Erase: Erases by pointing.
.: By typing a period ".", a straight line can be drawn from the end of the sketched line.
Pen: Pen up/down. By pressing "P," the pen will be raised up if it is down and lowered if it is up. This is identical to using the "pick" or pointing button on the pointing device.
Record: Entering "R" will record the lines already done.
Connect: While the pen is up, entering "C" and moving the crosshairs to the last sketch will connect the next freehand line to the last one sketched.

Drawing a Table and Chairs

You will draw a table and chair using a number of edit commands to make the task easy. The illustration shows what the table and chair should look like in the next sequence (Fig. 5-3).

1. Turn on ORTHO on status line or use F8 key.
2. Turn on SNAP on status line or use F9 key.
3. PAN over using scroll bars to draw chair and table.
4. Use LINE command to draw a table 3 feet square (using C to close polygon). If coordinate tracking is not working, press F6 key twice.
5. Change SNAP to 1" and make sure it is on.
6. Draw a side chair 18" square with a 2" back.

Tip: You can use object snap to make sure chair is centered in front of table. Use the MOVE command (making sure you select all lines on the chair). Use midpoint of front of chair for first point of displacement and then use midpoint on the front of the table for the second point of displacement. Then, move chair a little ways from the table.

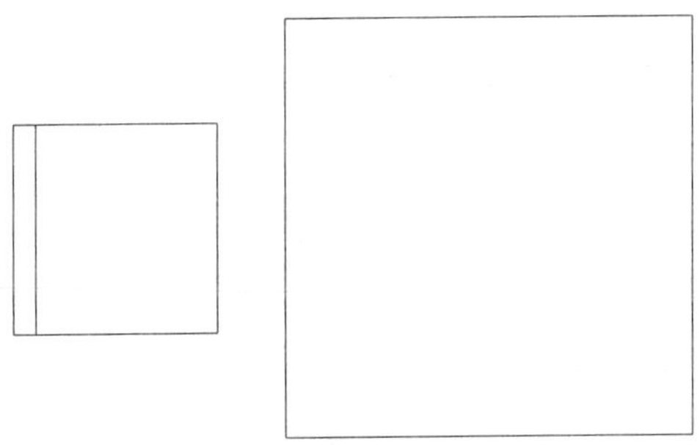

Fig. 5-3 Draw a square table and chair as shown.

FILLET Command

This command is used to round the corners of two intersecting lines. With the radius at 0 or the default value, you can use the FILLET command to make two lines intersect. To round off the corners of the chair, use the FILLET command with radius set at 2" and refer to illustration (Fig. 5-4).

1. **Turn SNAP off by using F9 key or by clicking on the status line.**

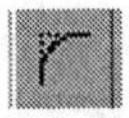

Command: **FILLET**
Current settings: Mode = TRIM, Radius = 0'-0 1/2"
Select first object or [Polyline/Radius/Trim]: **R**
Specify fillet radius <0'-0 1/2">: **2"**
Command: **(press ENTER)**
FILLET
Current settings: Mode = TRIM, Radius = 0'-2"
Select first object or [Polyline/Radius/Trim]: **(point A)**
Select second object: **(point B)**
Command: **(press ENTER)**
FILLET
Current settings: Mode = TRIM, Radius = 0'-2"
Select first object or [Polyline/Radius/Trim]: **(repeat for other corners**)

The chair should appear as in the illustration (Fig. 5-5).

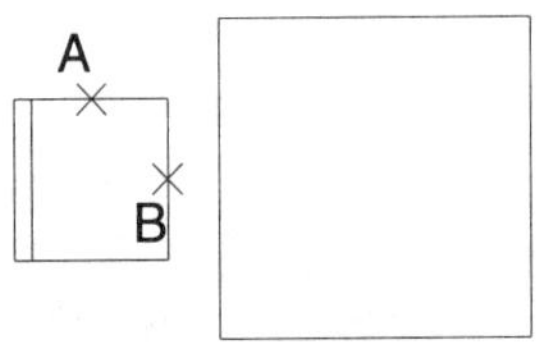

Fig. 5-4 Selection points for FILLET command.

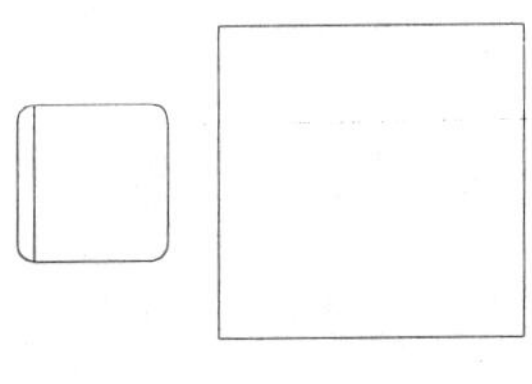

Fig. 5-5 Chair after using FILLET command on all corners.

You are going to have four chairs around this table so you will use the ROTATE command to rotate one chair.

1. **If needed, ZOOM out so that you can place all three chairs.**
2. **Make sure ORTHO is on (F8 key) by checking the status line.**

You will make a copy of the chair and move the copy to above the chair and away from the table (Fig. 5-6). To round the corner, the FILLET command makes a series of arcs. Thus it is important when moving or copying this chair in the next sequence to make a window around the chair to make sure you select the arcs.

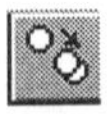

Command: **COPY**
Select objects: **(put a window around just the chair)**
Select objects: **(press ENTER or right click)**
Specify base point or displacement, or [Multiple]: **(point on chair)**
Specify second point of displacement or <use first point as displacement>:
(move to above table as in Fig. 5-6)

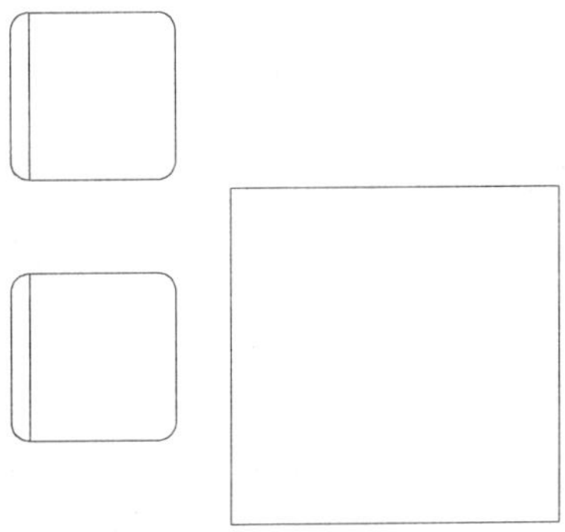

Fig. 5-6 Copy the chair and move copy to the top.

ROTATE Command

The ROTATE command is used to rotate objects around an axis you select. You will rotate the copy of the chair 90 degrees (Fig. 5-7).

Command: **ROTATE**
Current positive angle in UCS: ANGDIR=counterclockwise ANGBASE=0
Select objects: **(use a window to select the chair)**
Select objects: **(press ENTER or right click)**
Specify base point: **(point A)**
Specify rotation angle or [Reference]: **-90**

The result should look like the illustration (Fig. 5-8).

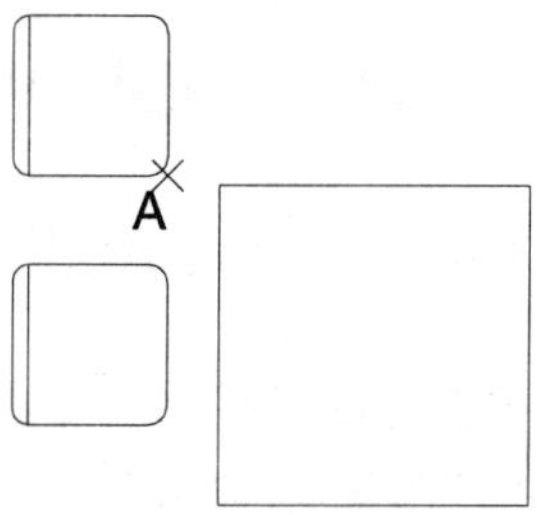

Fig. 5-7 Base point to rotate chair.

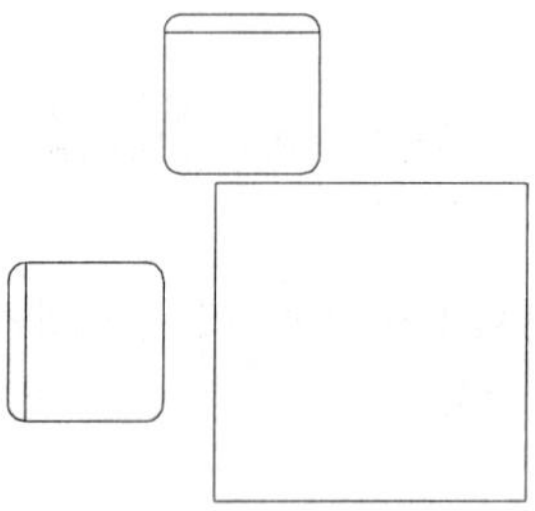

Fig. 5-8 Chair after rotating.

MIRROR Command

The MIRROR command makes a mirror image or copy. You will use this command to make the chairs for the opposite side of the table. Make sure your table is square or else this will not work correctly. Follow the illustration for the points (Fig. 5-9).

1. Use the MOVE command to move the chair as illustrated (Fig. 5-9).
2. Turn ORTHO off by using the F8 key or clicking on ORTHO on status line.

Command: **MIRROR**
Select objects: **(make a window around one chair)**
Select objects: **(make a window around the other chair)**
Select objects: **(press ENTER or right click)**
Specify first point of mirror line: **int (or click on** **)**
of **(point A)**
Specify second point of mirror line: **int (or click on** **)**
of **(point B)**
Delete source objects? [Yes/No] <N>: **(press ENTER)**

If you had responded "Y" to "Delete old objects?" AutoCAD would have deleted the original items leaving you with the mirrored copy only.

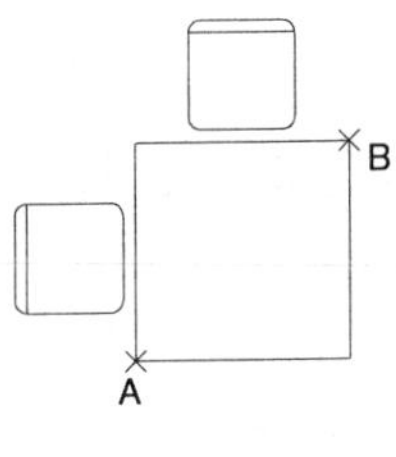

Fig. 5-9 Points for mirror line.

The result should be as in the illustration (Fig. 5-10).

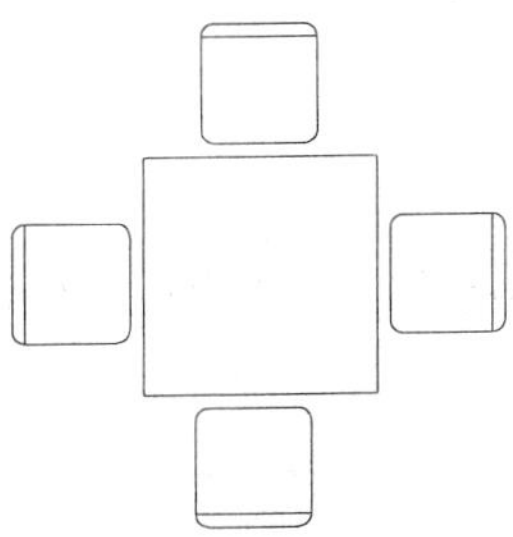

Fig. 5-10 Result after mirroring the chairs.

SCALE Command

The SCALE command is used to change the size of objects. Furniture is typically not scaled since it is drawn to full scale. Types of objects that might be scaled are logos, plants, or trees.

In the next sequence, you will first draw a plant although it need not look like the illustration (Fig. 5-11). This plant was drawn with SNAP at 2" using both the CIRCLE and LINE commands with use of object snap (cen, int, nea) to connect the lines precisely. One branch was drawn, the ARRAY command was used to make a polar array.

1. Draw a plant using circle and other draw commands.
2. Use COPY command to make a copy of the plant.

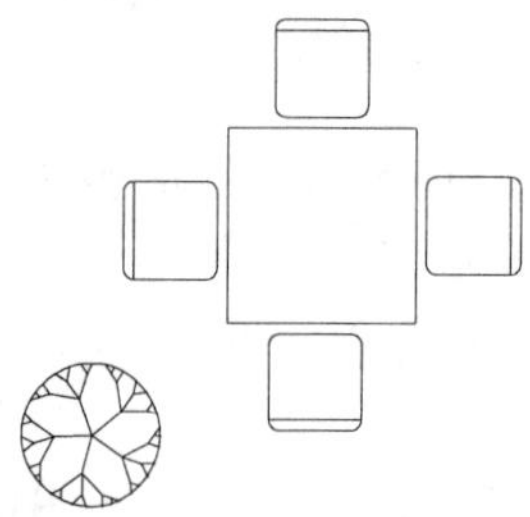

Fig. 5-11 Draw a plant using circles, lines or arcs.

Command: **SCALE**
Select object: **(select one plant with a window)**
Select objects: **(press ENTER or right click)**
Specify base point: **cen (or click on 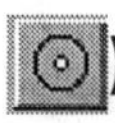)**
of **(point on circle circumference)**
Specify scale factor or [Reference]: **.8**

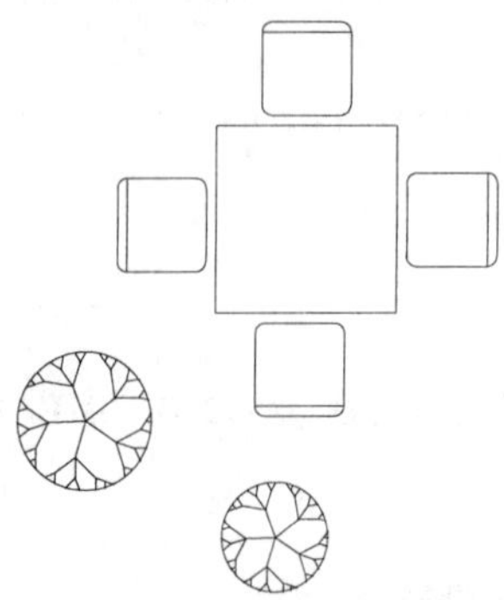

Fig. 5-12 Copied plant is scaled to make smaller plant.

You could also specify a scale factor of 2 to enlarge it to twice its size, .5 to decrease it 50%, etc.

SHORT ANSWER/DISCUSSION QUESTIONS

1. What command do you use to reduce the size of any object?__________

2. Why should you not scale down a drawing such as a floor plan, elevation, etc.?

3. Why is the SKETCH command not recommended for extensive use on drawings?

4. What command makes solid circles?

5. What two commands make solid rectangular shapes?

PRACTICE

1. Make a two-dimensional design on letter-size paper (8-1/2" x 11") with overlapping squares, circles, and rectangles. Use BHATCH in various colors to make a pleasing design (unity, rhythmn, etc.).

ADDITIONAL WORK

TRACE Command

The TRACE command is used to make a thick line. It is an old AutoCAD command--most users now use the PLINE.

Command: **TRACE**
Specify trace width <0'-0 1/16">: **6**
Specify start point: **(point)**
Specify start point: **(point)**
Specify start point: **(press ENTER or right click)**

You have to enter the third point before you see the thick line.

FILL Command

If solids do not appear solid, FILL mode or command is off. To turn on:

Command: **FILL**
Enter mode [ON/OFF] <ON>: **ON**

Part Three: Drafting

TUTORIAL 6

DRAFTING AN OFFICE PLAN

Commands Learned:	MLINE	VIEWRES
	DIST	BLOCK
	MLEDIT	INSERT

This tutorial will introduce you to drafting with AutoCAD including use of drawing symbols. In addition, you will learn more about commands presented earlier.

1. Start a new drawing and name it with tut6 and your initials.

Set up your drawing for C size (18" x 24"), 1/4"=1' scale paper:

UNITS	**Architectural**
LIMITS	**upper right corner 96',72'**
GRID	**5'**
ZOOMall	

You are going to be drafting the office illustrated below (Fig. 6-1). The building has outer walls of masonry that are one foot thick and interior movable partition walls of four inches. The tutor will take you through the process step by step. At the same time, you will be learning about more drawing, edit, and display aid commands. Try not to skip any part of this process or follow steps out of sequence. This file will be used for the next four exercises also.

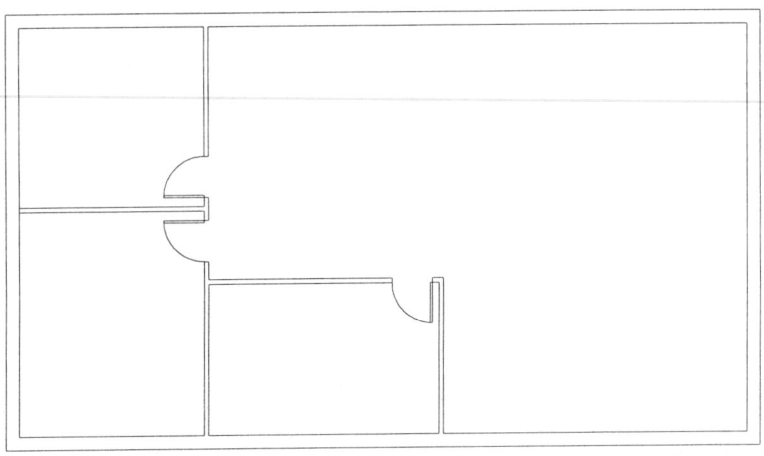

Fig. 6-1 Floor plan at end of this tutorial.

There are several ways you could draw the walls. The simplest woud be just to use the PLINE command, either entering the coordinate pairs, or using SNAP and ORTHO to just point. Then you could use the OFFSET command to make the interior wall--this is actually the most common method. Another method is to use the MLINE command which creates multiple lines.

MLINE Command

The MLINE command draws double lines, a task useful for drawing walls. You can also create your own multiple lines by creating them with the MLSTYLE command which is explained in the User's Guide. You will first change the scale of the multiple line by entering 12, which will give you a thickness of 12" or 1' to the outside wall. You will set the justifcation so that the walls are drawn from the middle of the wall. You will be entering the polar coordinates. Click on the icon on the draw menu or type:

Command: **MLINE**
Current settings: Justification = Top, Scale = 1.00, Style = STANDARD
Specify start point or [Justification/Scale/STyle]: **J**
Enter justification type [Top/Zero/Bottom] <top>: **Z**
Current settings: Justification = Zero, Scale = 1.00, Style = STANDARD
Specify start point or [Justification/Scale/STyle]: **S**
Enter mline scale <1.00>: **12**
Current settings: Justification = Zero, Scale = 12.00, Style = STANDARD
Specify start point or [Justification/Scale/STyle]: **60'6",8'6"**
Specify next point: **@56'<180**
Specify next point or [Undo]: **@31'<90**
Specify next point or [Close/Undo]: **@56'<0**
Specify next point or [Close/Undo]: **C**

Your walls should appear as in the illustration (Fig. 6-2).

Fig. 6-2 Exterior masonry walls drawn with MLINE command.

DIST Command

The DIST command will measure distances. You will use it to check on the horizontal dimension.

> **1. Turn ORTHO on (F8 key and check that ORTHO is on status line).**
> **2. Set SNAP to 6".**

Click on the "Distance" icon on the Standard toolbar at the top of the screen or type (Fig. 6-3):

Command: **DIST**
First point: **(point on line at point A)**
Second point: **(point on other line at point B)**
Distance = 55'-0", Angle in XY Plane = 0, Angle from XY Plane = 0
Delta X = 55'-0", Delta Y = 0'-0", Delta Z = 0'-0"

Fig. 6-3 Checking the interior width with the DIST command.

Check that AutoCAD responded with 55' which should be the horizontal dimension.

> **1. Use the DIST command to check that the inside vertical dimension is 30 feet.**

Turn off UCSICON command which turns off the icon pictured in the lower left corner. You will use it in later exercises:

Command: **UCSICON**
ON/OFF/All/Noorigin/ORigin<ON>: **off**

The next task is to add the two interior walls which are 4 inches thick.

1. **Set SNAP to 2".**
2. **Check status line to make sure ORTHO and SNAP are on.**
3. **Zoom to the drawing extents by using [icon] or by entering:**

Command: **Z**
ZOOM All/Center/Dynamic/Extents/Left/Previous/Vmax/Window/<Scale(X/XP)>:
E

Command: **MLINE**
Current settings: Justification = Zero, Scale = 12.00, Style = STANDARD
Specify start point or [Justification/Scale/STyle]: **S**
Enter mline scale <12.00>: **4**
Current settings: Justification = Zero, Scale = 4.00, Style = STANDARD
Specify start point or [Justification/Scale/STyle]: **19',9'**
Specify next point: **@30'<90**
Specify next point or [Undo]: **(press ENTER)**

Your plan should appear as in the illustration (Fig. 6-4).

Fig. 6-4 Drawing the interior walls with MLINE.

Next you will use the MLINE command to draw another wall to make the two offices on the left side of the plan:

Command: **MLINE**
Current settings: Justification = Zero, Scale = 4.00, Style = STANDARD
Specify start point or [Justification/Scale/STyle]: **5',25'8"**

Specify next point: **@13'10"<0**
Specify next point or [Undo]: **(press ENTER)**

Your plan should appear as in the illustration (Fig. 6-5). Note that you could have just pointed to those coordinate locations lining up the crosshairs with the walls because you had SNAP set to 2" increments.

Fig. 6-5 The second interior wall.

You will draw a door. To do so, you will zoom in the area where you want the door. The illustration shows the window you want to make with the ZOOM command (Fig. 6-6):

Command: **Z**
ZOOM
All/Center/Dynamic/Extents/Left/Previous/Vmax/Window/<Scale (X/XP)>:
(make a window as illustrated in Fig. 6-6)

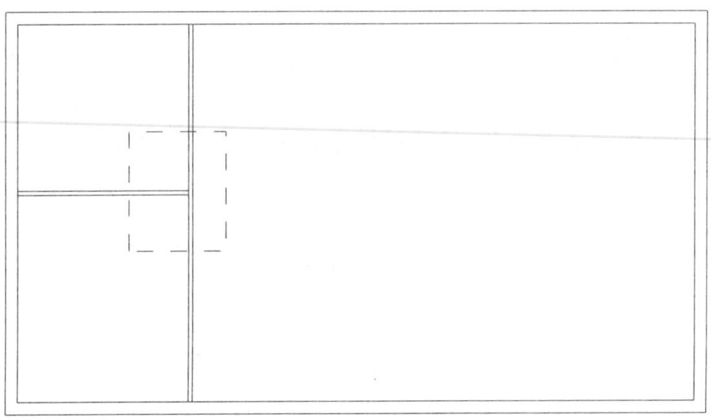

Fig. 6-6 ZOOM with a window to draw door.

You will draw the door and move it to the right location.

1. Make sure ORTHO is on.
2. Press COORDS key (F6) twice to enable coordinate tracking.
2. Set SNAP to 2".
3. With LINE command, draw a 3' wide door 2" thick (Fig. 6-7).

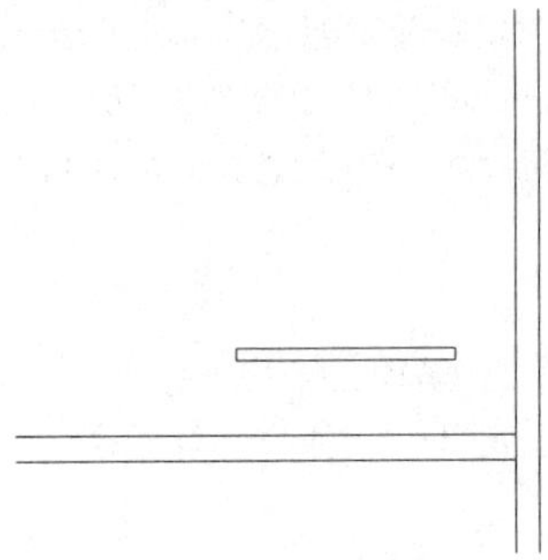

Fig. 6-7 Door.

Now draw an arc using the “center, start, and endpoint” method (Fig. 6-8). After you enter the center point, you should check coordinate tracking to make sure the arc has a 3' radius.

Command: **ARC**
Center/<Start point>: **C**
Center: **(point A and note Y coordinate)**
Start point: **(add 3' to Y coordinate and then point B)**
Angle/Length of chord/<End point> **(point C)**

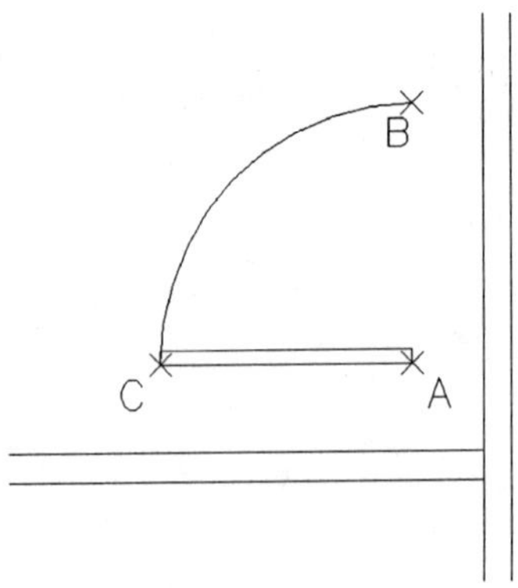

Fig. 6-8 Door swing.

Your door swing may not appear as in the illustration because the resolution of the screen is low. You will use the VIEWRES command to increase screen resolution of arcs and circles.

VIEWRES Command

This command controls the resolution of the arcs and circles on the display screen. With lower resolution, arcs are composed of only a few lines while higher resolution forces the arcs to be drawn with many lines resulting in a smoother appearance but slower regeneration. The VIEWRES command does not affect how circles and arcs are plotted; that is dependent upon the resolution of the printer or plotter. Answer yes when prompted for fast zooms because AutoCAD will use the fastest way to regenerate screen displays.

> Command: **VIEWRES**
> Do you want fast zooms? <Y> **(press ENTER)**
> Enter circle zoom percent (1-20000) <100>: **1000**

The drawing regenerates and the arc should appear smoother.

You will be using the door again in this floor plan so you will be introduced to the BLOCK command that will allow you to store the door in the drawing file.

BLOCK Command

The BLOCK command stores objects for later retrieval with the INSERT command. When asked for the insertion base point, you will use the corner of the door or where it would be hinged (Fig. 6-9).

> Command: **BLOCK**

A dialogue window appears (Fig. 6-9):

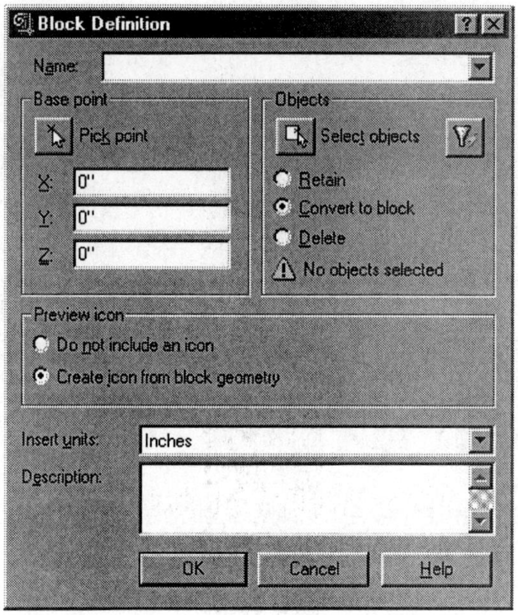

Fig. 6-9 Block Definition window.

1. Type "door" in Name box.
2. Click on "Pick point" and use "int" object snap to select Point A or lower right corner of door (Fig. 6-10).
3. Click on "Select objects" button.
4. Make a window around door and swing to select all lines and arc.
5. Press Enter key or right click to end selection.
6. Click "OK" to exit Block Definition dialogue window.

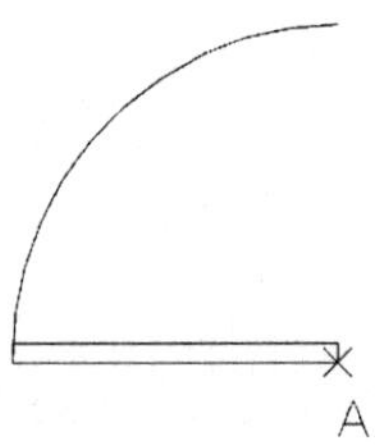

Fig. 6-10 Base point for insertion of door.

The door is converted to a block so that when you select by pointing on the door, the whole door and swing will be selected rather than only the individual lines or arc.

MLEDIT Command

You cannot use TRIM, BREAK, or other EDIT commands to edit the multiple lines. One way to use these edit commands would be to use the EXPLODE command (explained later) to revert the multiple lines to regular lines if they were done with the LINE command. But, MLINE has its own edit command, MLEDIT, which may be easier to use to break out walls or trim the multiple lines. You will use the MLEDIT command to break out the walls for the door:

1. Check status line to be sure ORTHO and SNAP are on.

1. Use MOVE command to move the door to the wall about 6 inches from corner of the room (Fig. 6-11).

The MLEDIT icon is found on the Modify 2 toolbar or type:

Command: **MLEDIT**

1. When the dialogue box appears, click on the fourth icon  from the left, middle row. It should say "Cut All" at the bottom after you click on it. Follow points (Fig. 6-11):

Select mline: **(point A)**
Select second POINT: **(point B)**
Select mline (or Undo): **(press ENTER or right click)**

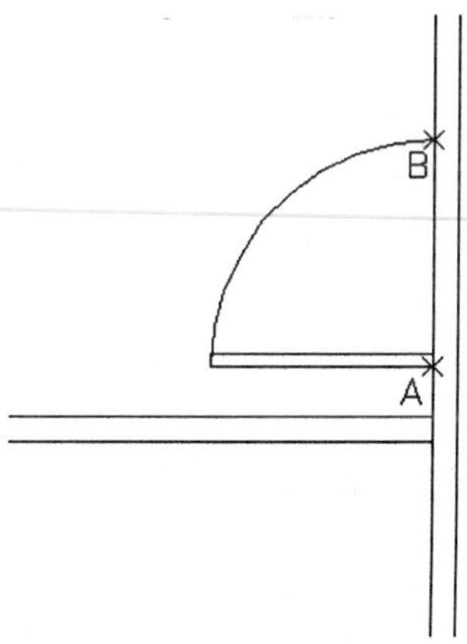

Fig. 6-11 Use of MLEDIT to break out walls.

1. Draw two lines to finish the door opening as in illustration (Fig. 6-12).

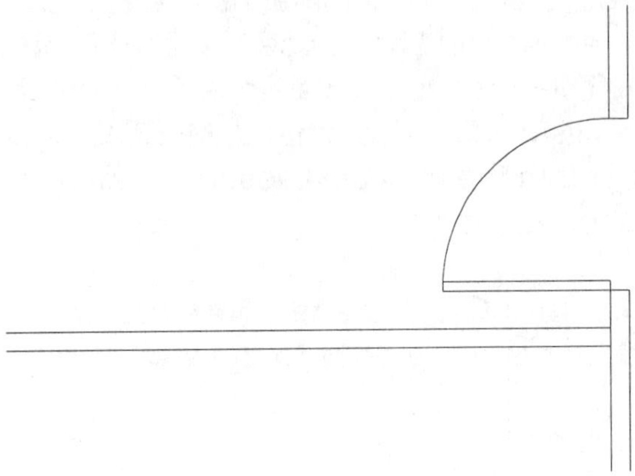

Fig. 6-12 Draw two lines to finish door opening.

Now you will use the MIRROR command to copy the door and two small lines to the side. You may need to toggle SNAP off when you select the lines but make sure you turn it back on for the mirror line. For the mirror line points, follow the illustration (Fig. 6-13):

1. Make sure ORTHO and SNAP at 2" are on.

Use the mirror icon on the Modify toolbar or type:

Command: **MIRROR**
Select objects: **(select door)**
Select objects: **(select the two 4" lines connected to door)**
Select objects: **(press ENTER or right click)**
First point of mirror line: **(point A)**
Second point: **(point B)**
Delete old objects? <N> **(press ENTER or right click)**

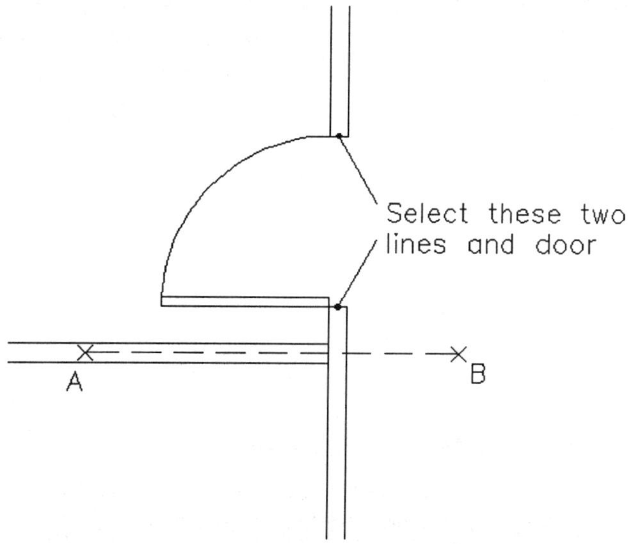

Fig. 6-13 MIRROR line

1. Use the MLEDIT command to open or break out the wall as you did before.
2. On menu bar, click on "View" and "Redraw" or type R to the command prompt to redraw the screen.

The result should be as in the illustration (Fig. 6-14).

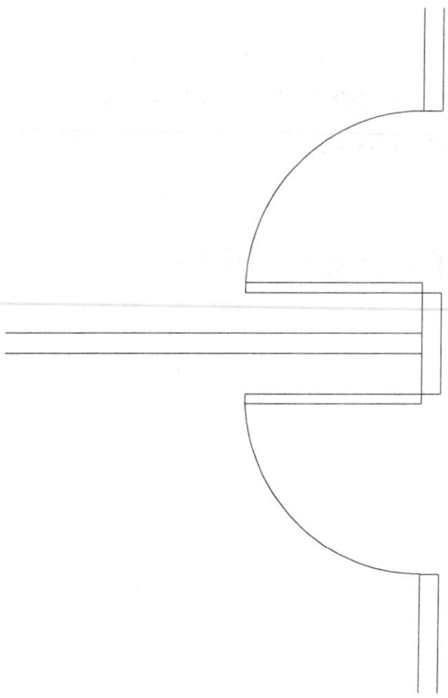

Fig. 6-14 After use of MIRROR command.

Next, you will use an alternate method of drawing walls, the PLINE and OFFSET commands combined with a couple of new OSNAP options to make the third office. Many designers prefer this method. The first OSNAP option is to use a temporary tracking point. Think of it as telling AutoCAD you want to start a certain distance from a point. The other OSNAP option is to make a line connect precisely to a line that is perpendicular to the line you are drawing.

1. Zoom to drawing extents by using icon or type Z and then E.

Refer to points in illustration (Fig. 6-15).

Command: **PLINE**
Specify start point: **FROM (or click on icon)**
Base point: **(point A to select the point you want to snap from)**
<Offset>: **@1'6"<270**
Current line-width is 0'-0"
Specify next point or [Arc/Close/Halfwidth/Length/Undo/Width]: **@17'6"<0**
Specify next point or [Arc/Close/Halfwidth/Length/Undo/Width]: **PERP (or click on icon)**
to **(point B)**
Specify next point or [Arc/Close/Halfwidth/Length/Undo/Width]: **(press ENTER or right click)**

1. Use OFFSET command to offset the polyline 4" inside.
2. Check with the DIST command that the inside dimension is 17'-2" in width.

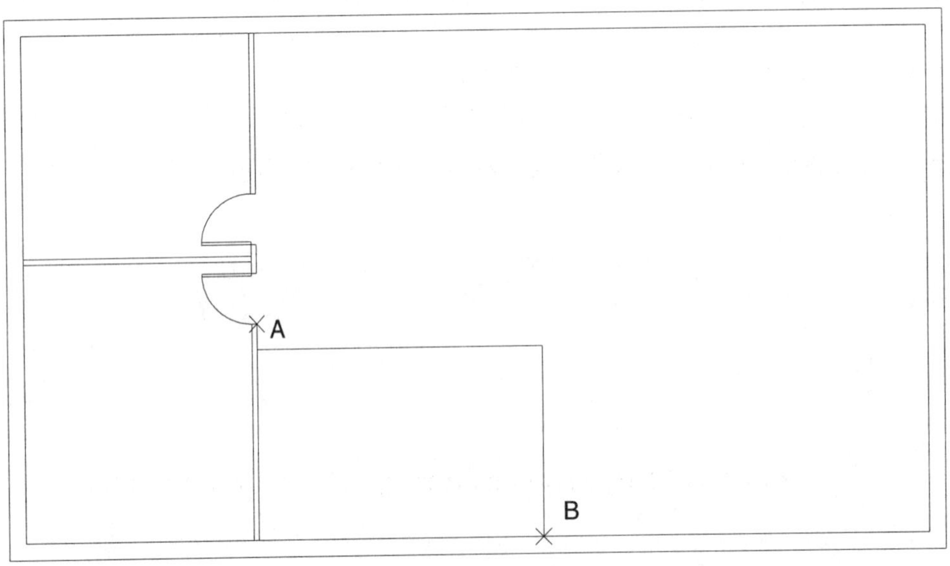

Fig. 6-15 Drawing the last office wall.

The or temporary tracking point is similar. It allows you to select a point to track from but instead of typing in polar coordinates, you can drag the cursor to the location with AutoCAD indicating the distance. Note that the Polar option on the status line should be off for this to work. Snap from, temporary tracking point, and perpendicular object snap are very useful when drafting walls.

EXPLODE Command

The EXPLODE command is used to convert polylines to regular lines or as if drawn by the LINE command. Then, you can use edit commands such as TRIM or BREAK. This command is also useful to explode hatch back to individual lines. Later you will learn more about this command but for now, you will use this command to explode the two polylines you just drew.

1. **Make sure ORTHO and SNAP (2") are on.**
2. **Zoom to upper right corner of the third office (Fig. 6-16).**

Command: **EXPLODE**
Select objects: **(select first polyline)**
Select objects: **(select second polyline)**

Command: **L**
LINE Specify first point: **TT (or click on icon)**
Specify temporary OTRACK point: **(point A)**
Specify first point: **(move crosshair to left until it reads "Track Point: 0'6"<180 degrees" and point B)**
Specify next point or [Undo]: **(point C)**
Specify next point or [Undo]: **(press ENTER or right click)**
Select object to offset or <exit>: **(press ENTER or right click)**

Fig. 6-16 Draw line using temporary tracking marker.

Next, you will offset the line to for the 3' opening.

Command: **OFFSET**
Specify offset distance or [Through] <0'-4">: **3'**
Select object to offset or <exit>: **(select small line you just drew)**

Specify point on side to offset: **(point to left of line)**

Command: **TRIM**
Current settings: Projection=UCS Edge=None
Select cutting edges ...
Select objects: **(point A)**
Select objects: **(point B)**
Select objects: **(press ENTER or right click)**
Select object to trim or [Project/Edge/Undo]: **(point C)**
Select object to trim or [Project/Edge/Undo]: **(point D)**
Select object to trim or [Project/Edge/Undo]: **(press ENTER or right click)**

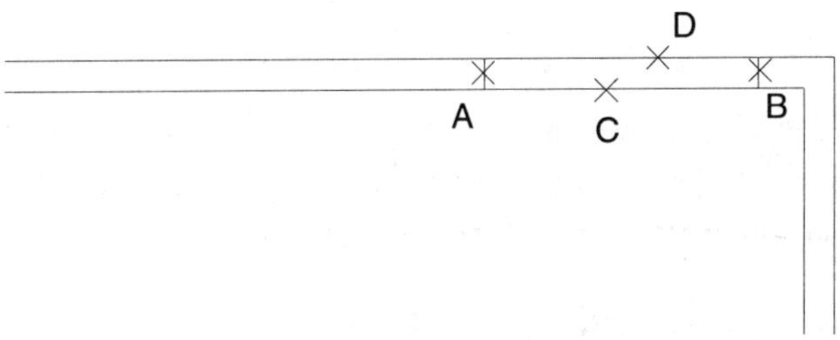

Fig. 6-17 Use the 2 small lines as cutting planes for TRIM command.

INSERT Command

Next you will insert the last interior door using the INSERT command which inserts blocks or drawing files:

Command: **INSERT**

The Insert dialogue window appears. Door should appear as the name.

1. **Enter 90 for angle.**
2. **Click on "OK" to exit dialogue window.**

3. Point to insert door in wall.
4. Draw two lines for the door opening again.
5. Use TRIM or BREAK to break out the walls (Fig. 6-18).

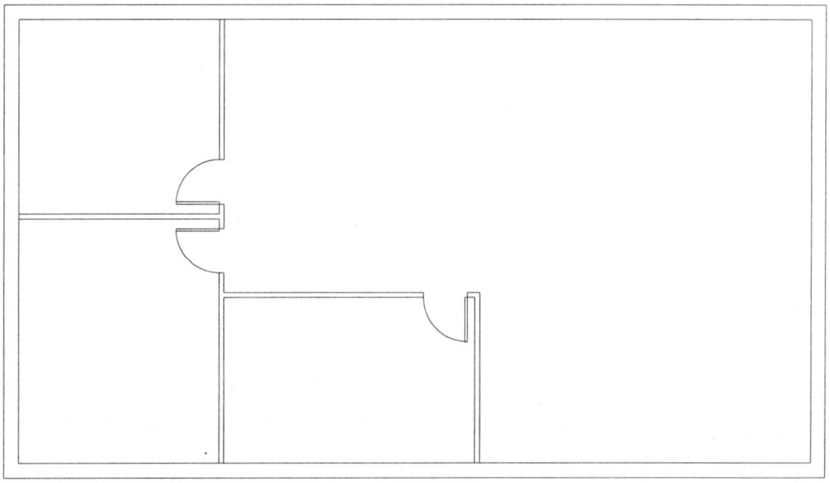

Fig. 6-18 Final plan.

This is the end of this tutorial. You will be using this drawing in the next several tutorials so be sure to save it. This office plan can be printed by using "extents" for area to plot and "fit" for scale if you are plotting to a printer.

SHORT ANSWER/DISCUSSION QUESTIONS

1. What command is used to measure distance in a drawing?_________

2. What command is used to edit lines made with the MLINE command?_______

3. What command is used to store symbols?__________

4. What command do you use to put those symbols into a drawing?__________

PRACTICE

1. If you have a project to work on, this is a good time to start practicing with AutoCAD to draft a floor plan with doors and door openings.

TUTORIAL 7

SPACE PLANNING THE OFFICE

Commands Learned:	LAYER	VIEW	ATTDISP
	PURGE	EXPLODE	ATTEDIT
	WBLOCK	ATTDEF	XREF
	MINSERT	ATTDIA	
	GROUP		

This tutorial will introduce you to space planning with AutoCAD with more use of blocks for storage of furniture and use of layers to make drawings.

1. Menu Bar: File > Open.
2. When dialogue window appears, click on the drawing file you used in the last tutorial.
3. Click on "OK" to open file.

You will be drafting the furniture illustrated in the office plan (Fig. 7-1). The tutorial will guide you through each drawing.

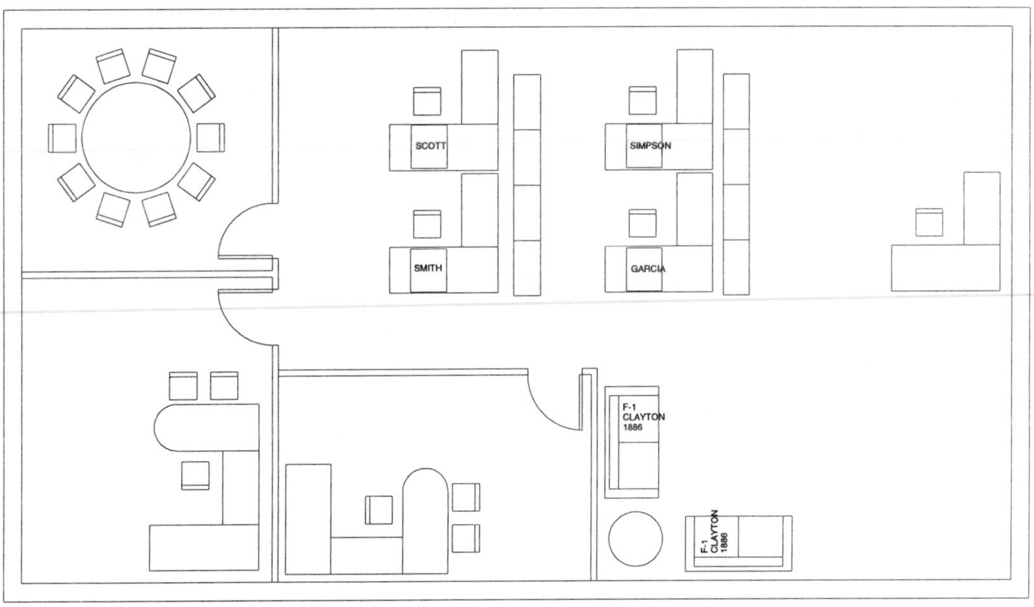

Fig. 7-1 Final office furniture plan.

On furniture that is sized in increments of 3", 6", or such, it is easiest to count the dots of the 1' grid and use snap to facilitate drafting. You will want to use the "C" option of LINE command to close the polygon when drawing the file and make sure that coordinate tracking is on (press F6 key twice if it is not tracking when you are in the LINE command). Now draft the lateral file:

1. Zoom with a window to a small area above the floor plan to draw the furniture.
2. Change SNAP to 3" and make sure it is on.
3. Check on status line that ORTHO is on.
4. Make sure the GRID is on (F7 key).
5. Use the LINE command to draw a lateral file 3' wide by 1'-6" deep as illustrated (Fig. 7-2).

Now you will store the lateral file as a block. When asked for the insertion base point, you will specify the lower left corner as this is a typical way to store furniture blocks (Fig. 7-3).

Command: **BLOCK**

1. Type "file" in Name box.
2. Click on "Pick point" and use "int" object snap to select point A.
3. Click on "Select Objects" button.
4. Select all 4 lines and press Enter key or right click to end selection.
5. Click "Ok" to exit Block Definition dialogue window.

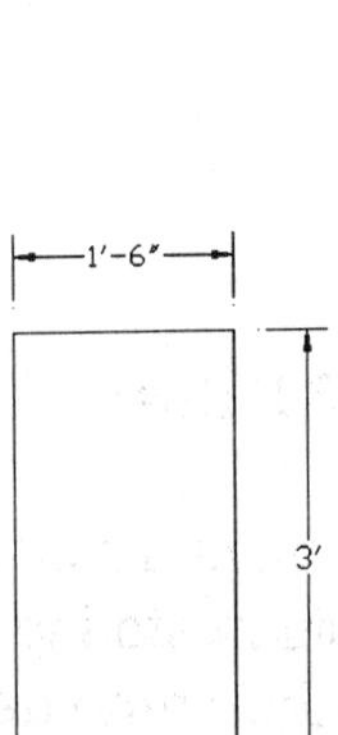

Fig. 7-2 Lateral File

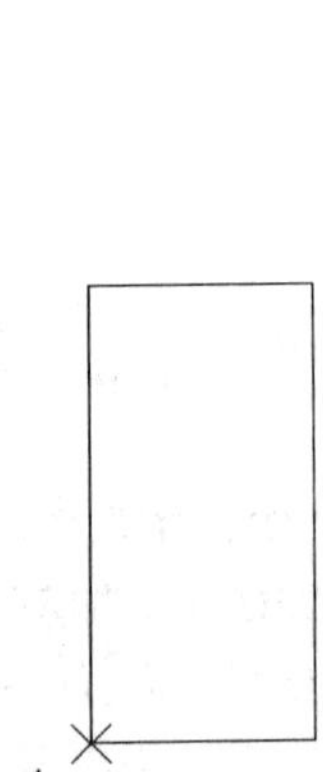

Fig. 7-3 Insertion Point

Now you will draft the desk referring to the illustration (Fig. 7-4):

1. Check that ORTHO and SNAP (3") are on.

2. Use LINE command to draft a desk 6' wide by 2'-6" deep.

Using the BLOCK command again, name the block "desk" and use the lower left corner of the desk for the insertion base point (Fig. 6-18):

Command: **BLOCK**

1. Type "desk" in Name box.
2. Click on "Pick point" and use "int" object snap to select point A.
3. Click on "Select Objects" button.
4. Select all 4 lines and press Enter key or right click to end selection.
5. Click "Ok" to exit Block Definition dialogue window.

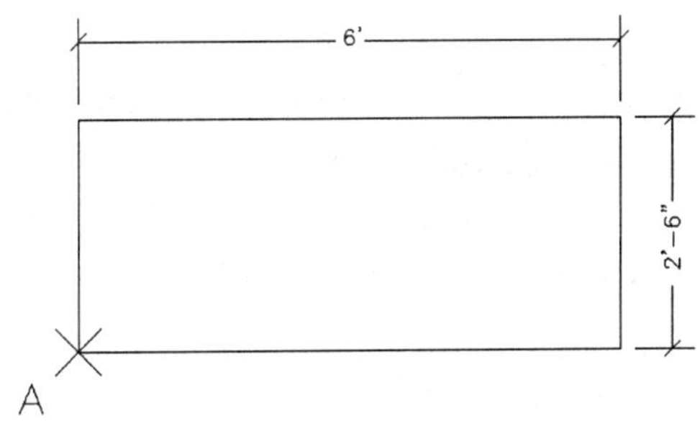

Fig. 7-4 Draw a 72" x 30" desk.

If you made a window around the desk, you may have slipped and pointed to the other corner right next to the first one which results in storing an empty space as the block. If so, repeat the BLOCK command, enter the name desk. AutoCAD will respond that this block already exists and will ask if you want to redefine it. Enter yes when AutoCAD asked if you want to redefine the desk.

Next, you will draft the return and store it as a block (Fig. 7-5):

1. Use the LINE command to draft the return 4' wide by 2' deep (Fig. 7-5).

Command: **BLOCK**

1. Type "return" in Name box.
2. Click "Pick point" using "int" object snap to select point A (Fig. 7-5).
3. Click "Select Objects" button.
4. Select all 4 lines and press Enter key or right click to end selection.
5. Click "Ok" to exit Block Definition dialogue window.

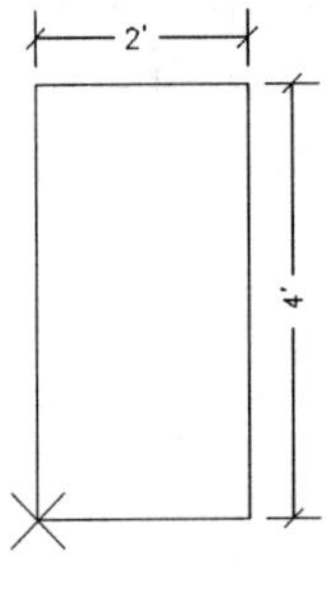

Fig. 7-5 Return.

Next, you will draft the bullet-shaped table (Fig. 7-6) and store as a block (Fig. 7-7):

1. Use LINE command to draft 3 lines as illustrated (Fig. 7-6).

2. Use the FILLET command and use the parallel lines for selecting the objects for the FILLET to make the arc.

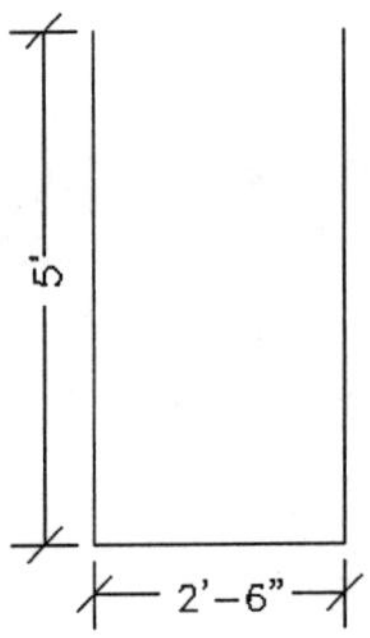

Fig. 7- 6 Three lines for bullet table.

Command: **BLOCK**

1. **Type "btable" in Name box.**
2. **Click "Pick point" using "int" object snap to select point A (Fig. 7-7).**
3. **Click "Select Objects" button.**
4. **Select all 4 lines and press Enter key or right click to end selection.**
5. **Click "Ok" to exit Block Definition dialogue window.**

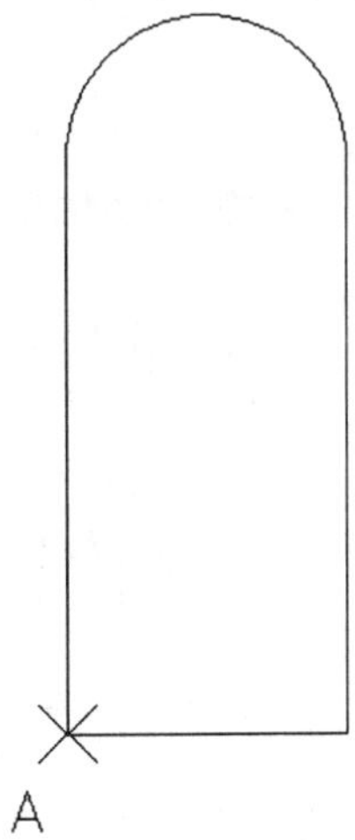

Fig. 7-7 Filleted lines and insertion point.

Finally, you will draft the task chair:

1. **Change SNAP to 3"**
2. **Use LINE command to draft chair, 18" square with 3" chair back.**

Using Commands Transparently

It is possible to use a command while you are still in the middle of a command sequence. This is called using a command transparently. You do so by typing a single apostrophe before the command name. Some of the icons such as the realtime zoom also invoke a transparent command. You will use SNAP transparently to change it to 1" (Fig. 7-8):

Command: **BLOCK**

1. **Type "chair" in Name box.**
2. **Click on "Pick point".**

Specify insertion base point: **'SNAP**
>>Snap spacing or ON/OFF/Aspect/Rotate/Style <3"> **1"**

3. **Select center of chair or Point A (Fig. 7-8).**
4. **Click on "Select Objects" button.**
5. **Select all 4 lines and press Enter key or right click to end selection.**
6. **Click "Ok" to exit Block Definition dialogue window.**

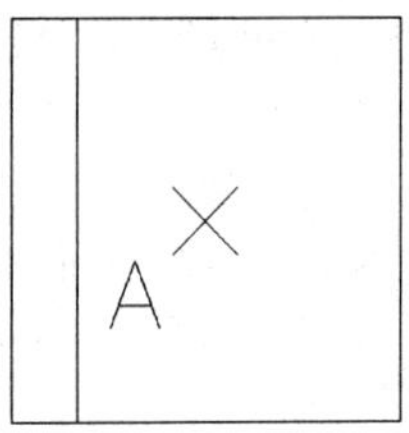

Fig. 7-8 Using a transparent command while blocking the chair.

You should now have six blocks. To check that you do, you will use another way to enter the BLOCK command

1. **Menu Bar: Draw > Block > Make...**
2. **Click on arrow down button in Name box to see block names.**
3. **Click on "Cancel" to exit "Block Definition" window.**

LAYER Command

The layering feature of AutoCAD allows you to create several drawings in one drawing file. You then can "turn off" what you do not need for printing or plotting out. Think of the layering feature as layers of tracing paper, each having a different drawing over the bottom tracing paper. "Turning off" the layer is analogous to removing the tracing layer. Use the layer icon or type:

Command: **LA**

The Layer Properties Manager window appears. Notice there is only one layer listed--layer 0. This layer has special properties and its name cannot be changed from 0. You will make a new layer "A-FURN", assign the color magenta, and make it the current drawing layer:

1. **Click on "New."**
2. **In the layer list, type "A-FURN" over the "Layer1" name.**
3. **Click on the white box under "Color" column. .**
4. **Under the "Standard Colors", click on the magenta or pink color.**
5. **Click on "OK" to exit the color window.**
6. **Check that the white box turned to magenta for the A-FURN layer.**
7. **While A-FURN is still highlighted or selected, click on "Current."**
8. **Click on "OK" to exit the Layer Properties Manager window.**

If you wanted to create a number of layers at one time, you would keep clicking on "New" until Layer3, Layer4, etc. appeared in the layer list. Layer names used in this book are the same as the layer names in the *CAD Layering Guidelines* published by the AIA (American Institute of Architects) and referenced in the last section of this book.

In the Object Properties toolbar, A-FURN should appear in the display next to the layer icon. All lines and drawing entities entered while the current layer is A-FURN are a part of that layer. You will learn more about the LAYER command in the next exercise. You will use the INSERT command to insert your furniture on the furniture layer. First, you will insert the desk (Fig. 7-9):

1. **Use ZOOM window**  **to make a window around plan (Fig. 7-9).**
2. **Change SNAP to 3" and check status line to make sure it is on.**
3. **Make sure ORTHO is on (F8 key).**

Command: **INSERT**

1. Click on desk in list of names.
2. Click "OK" button.
3. Place desk in approximate loction as illustrated (Fig. 7-9)

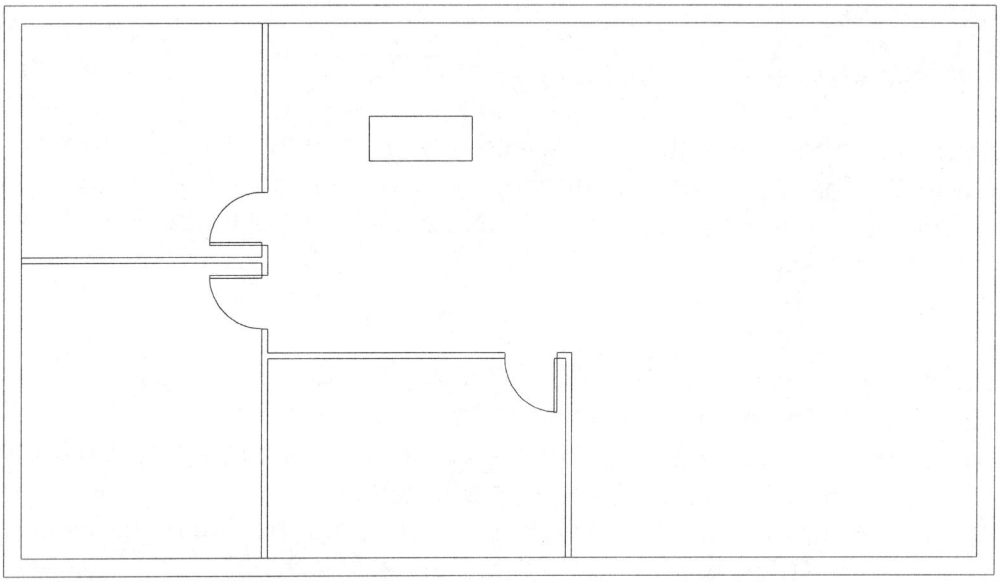

Fig. 7-9 Insert one desk.

The rotation angle works counterclockwise. If you had not stored your block in the same position as you needed it, you could enter a rotation angle. Keeping ORTHO on makes the rotation abruptly horizontal or vertical.

The X scale and Y scale factors allow you to change the scale of the block. While in this exercise, it does not make sense to change the scale of the furniture, there are many times that this feature will save you time. For instance, you may have drawn a plant or tree and stored it as a block. To make it twice as big, simply type 2 for both the X and Y scale factors. You could make a hedge three times as wide by typing 3 for the X scale factor and keeping 1 for the Y factor. In this way, you can use the X and Y scale factors to stretch logos or text.

1. Repeat INSERT command to place "return" block to line up with desk (Fig. 7-10).
2. Repeat INSERT command to place "chair" block (Fig. 7-10).
3. If needed, use the MOVE command  to align desk and return.

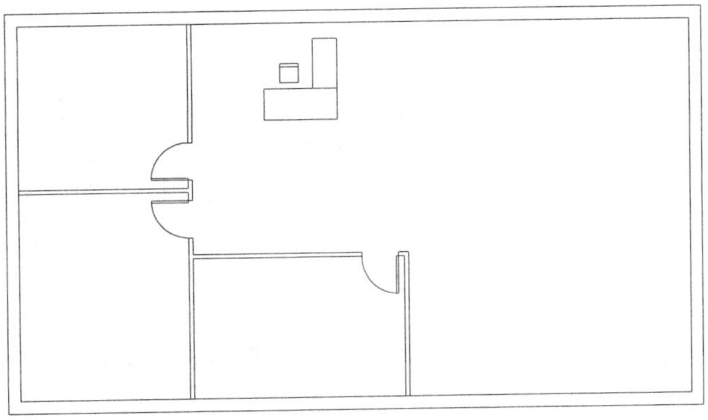

7-10 Insert return and chair.

1. Use the COPY command 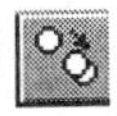 to make a copy of the desk, return, and chair and put these in front as in the illustration (Fig. 7-11).

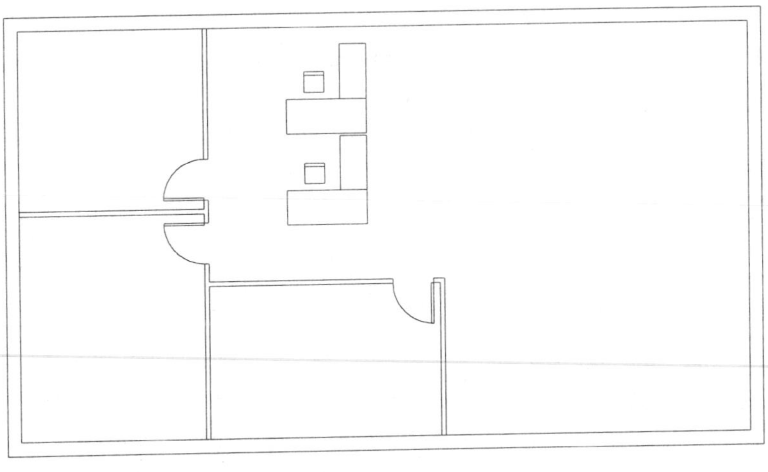

Fig. 7-11 Copy desk, return, and chair.

Notice that when you point to just one line of the object, the whole entity is selected. AutoCAD treats a block as a single entity, which has several advantages. First, moving and copying blocks are easier. Secondly, blocks will take up less space in the drawing file and so you will accumulate fewer bytes.

Next, you will insert the files using the MINSERT command.

MINSERT Command

The MINSERT command is similar to the rectangular array option of the ARRAY command. It allows you to make a rectangular array of blocks. A single block reference is created for the rectangular array which results again in reduced disk space and faster regeneration of the drawing. The single block reference has disadvantages, though, in editing. When moving the entities, for instance, you must move all of them and in the same configuration.

The following sequence assumes that the file did not have to be rotated. If you need to rotate the file 90 degrees, reverse the number of rows and columns (i.e. rows=1 and columns=5):

Command: **MINSERT**
Enter block name or [?]: **file**
Specify insertion point or [Scale/X/Y/Z/Rotate/PScale/PX/PY/PZ/PRotate]: **(point A)**
Enter X scale factor, specify opposite corner, or [Corner/XYZ]<1>: **(press ENTER)**
Enter Y scale factor <use X scale factor>: **(press ENTER)**
Specify rotation angle <0>: **(press ENTER)**
Enter number of rows (---) <1>: **4**
Enter number of columns (|||) <1>: **(press ENTER)**
Enter distance between rows or specify unit cell (---): **3'**

If the number of columns had been greater than one, you would have been prompted for the distance between columns.

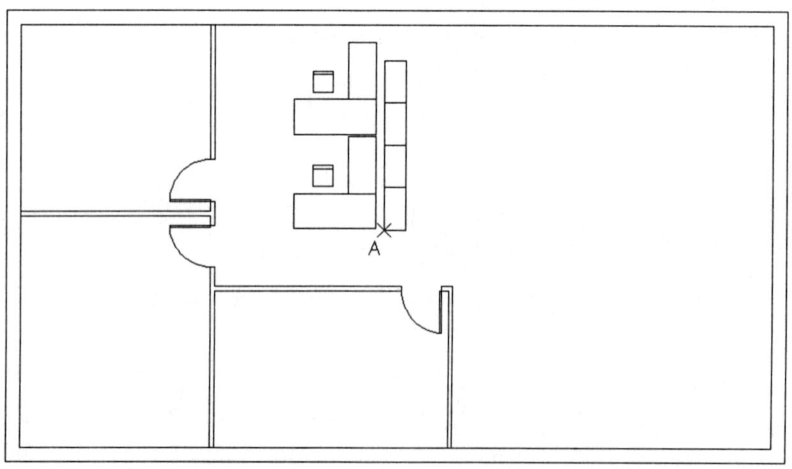

Fig. 7-12 Placing files with MINSERT command.

1. Use the COPY command to copy the furniture as illustrated (Fig. 7-13).

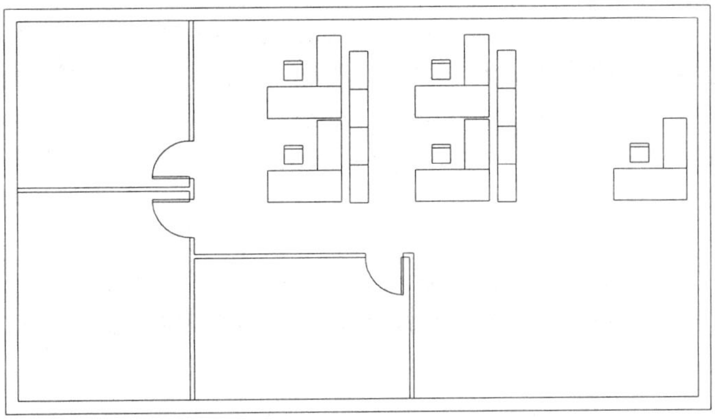

Fig.7-13 Copied desks.

VIEW Command

In space planning, you often have to zoom out to the whole drawing and then zoom to the same place in a drawing. The VIEW command is useful to save that view and then restore it when you need it. It is used most often for saving views when working in 3D. You can use the named views icon on the Standard Toolbar. First, zoom in to just see the plan:

1. Use ZOOM command to zoom into area as illustrated (Fig. 7-14).

2.Click on the named views icon or type VIEW at command prompt.

3. Click on "New" button.

4. Type in "vp" for view name.

5. Click on "OK" to exit the View window.

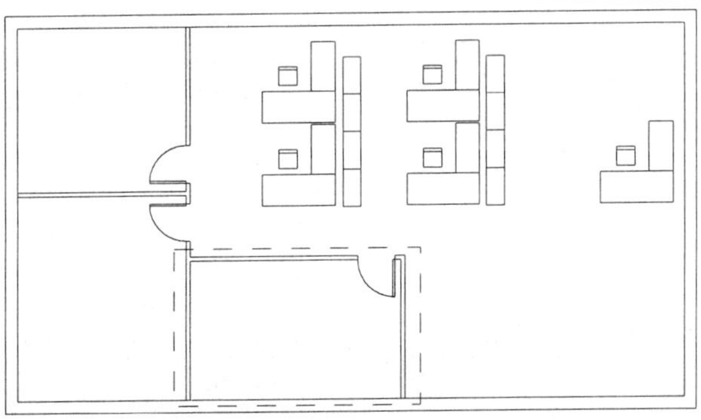

Fig. 7-14 Store view of office.

To restore it later, you would just use the named views icon, highlight vp, click on Restore, and OK to bring back the view of.the office.

1. Use the INSERT command to place the desk, return, btable, and chair blocks in the office (Fig. 7-15).

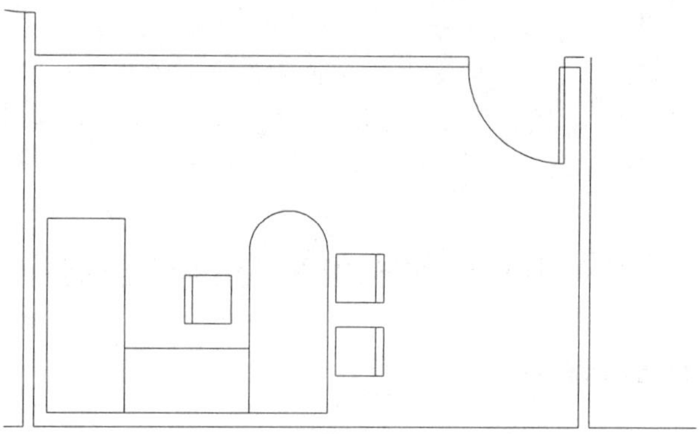

Fig. 7-15 Arrange furniture in VP's office.

About Nested Blocks

When you inserted the desks and chairs you made good use of the COPY command. However, space planning may be facilitated by using nested blocks. For instance, in a large building plan, you may have 30 or 40 desks with the exact same chair and components for each station. When using nested blocks, you must name the block with a name different from any of the block names in the nested block. The blocks can still be used separately, and you have the advantage of inserting both at once with the nested block.

On this plan, it makes sense to make a nested block of the furniture in the executive office because you will use the same furniture but rotated 90 degrees in the corner office (Fig. 7-16).

1. Use the BLOCK command and select the desk, return, btable, and chair blocks in the office (Fig. 7-16).
2. Name the block "exec."
3. Select point A for point of insertion (Fig. 7-16).

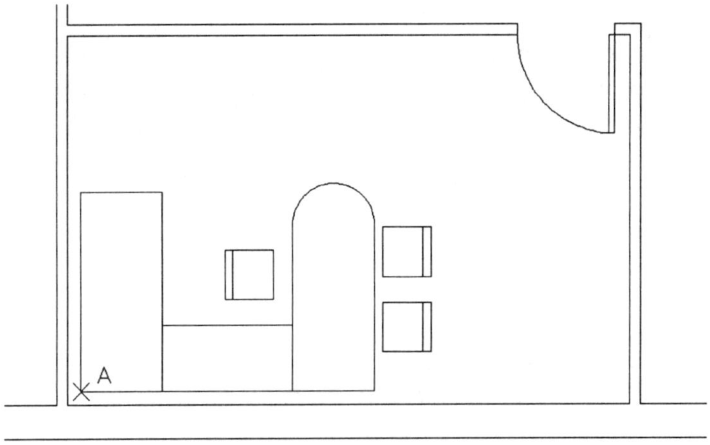

Fig. 7-16 Insertion point for nested block.

1. ZOOM to drawing extents .

2. ZOOM to just see office in lower left corner .

Command: **INSERT**

1. Select "exec" block.
2. Enter 90 degrees for angle.
3. Click "OK" to exit Insert dialogue window.

Specify insertion point or [Scale/X/Y/Z/Rotate/PScale/PX/PY/PZ/PRotate]: **(point A)**

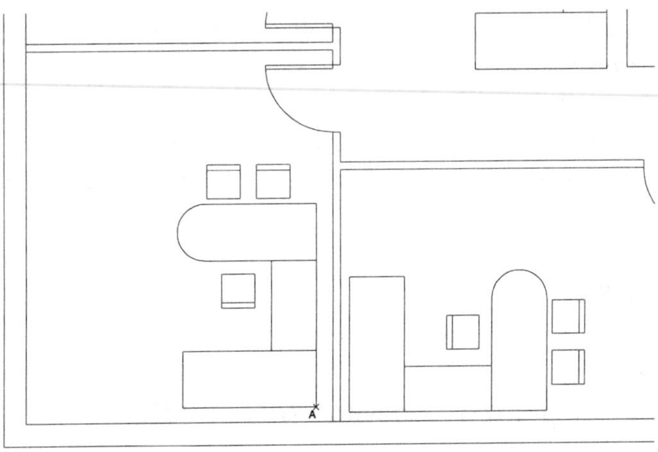

Fig. 7-17 Using a nested block.

GROUP Command

Nested blocks are useful when you want to use the same set of furniture. However, you might have situations when you want to group objects together temporarily and ungroup them and then make a different group. The GROUP command does just that. Just to see how it works, you will enter a group name for the furniture in the open office area.

Command: **GROUP**

1. **When dialogue window appears, type "clerical" in Group Name box.**
2. **Click on "New" button. When AutoCAD prompts:**

Select objects for grouping:
Select objects: **(Select the five desks, returns, chairs, and the eight files in open office area.)**
Select objects: **(press ENTER or right click)**

1. **Click on "OK" to exit "Object Grouping" dialogue window.**
2. **Use the MOVE command and click on one furniture item and try to move it.**

All the furniture is selected when you click on one. An object can be a member of more than one group. If you are unsure of which object is in a group, click on the group name in the "Object Grouping" dialogue window and then the "Highlight" button. AutoCAD will show you the objects in the group. The Explode button deletes the group definition. You will keep the "clerical" group, but turn the group off by turning off the PICKSTYLE system variable.

Command: **PICKSTYLE**
Enter new value for PICKSTYLE <1>: **0**

2. **Use the MOVE command again to move some desks.**

Now you can select just one desk.

WBLOCK Command

The BLOCK command creates blocks to be used within a drawing file. The WBLOCK command will write blocks to a separate drawing file so that you can use them in any drawing. This is how symbol libraries of architectural symbols are created. Most firms or schools have symbol libraries either created in-house or purchased from a third-party vendor. You will make a WBLOCK of the door:

Command: **WBLOCK**

1. **When dialogue box appears, type "door" for File name.**
2. **Click "Select objects" and select one of the doors.**
3. Use same insertion point as before.

AutoCAD stores this file named "door.dwg" in the same location as the drawing file. Or, you can choose another location. It can be inserted into any drawing. Never use the same block name you have opened the file with as AutoCAD will find duplicate file names.

Not only can written blocks be inserted into a drawing, but whole drawings can be inserted into another drawing. When AutoCAD prompts for the block name, enter the name of the drawing file. When AutoCAD finds that the block is not in the drawing file, it then searches for the file on disk.

Design Center

AutoCAD 2000 has a Design Center feature which allows you to insert blocks, text styles, layers, and other entities from another drawing into your current drawing.

1. **Click the Design Center icon on the Standard Properties toolbar.**
2. **Click the Load icon.**
3. **Open "Samples" folder and then "DesignCenter" folder.**
4. **Open "Home - Space Planner.dwg" to load file.**
5. **Click on "Blocks"**
6. **Scroll down and drag "file cabinet - letter" to the drawing area.**
7. **Click the x to close the DesignCenter window.**

The EXPLODE command, introduced in the previous tutorial is very useful to explode polylines, done with PLINE command, or multilines, done with MLINE command back to individual lines (as if you did them with the LINE command). Here, you will EXPLODE a block back to individual objects in order to edit the block.

Command: **EXPLODE**
Select objects: **(click on one of the returns in general office area)**
Select objects: (**press ENTER)**

1. Use the ERASE command and select just one of the lines in the return.

One line was selected instead of the whole block because the return is no longer a block.

1. Use the ERASE command and erase the return.

2. Use INSERT command to insert another return.

To update inserted blocks, you would make a new block out of the edited block. Then, type "-INSERT" and when prompted with "Enter block name [?]:" enter "oldblockname=new block name" to have all the inserted blocks updated. This is very useful if you have inserted about 50 doors and you needed to change them.

Note that if you type a hyphen (-) before any command, it will use prompts at the command line rather than using a dialogue window.

PURGE Command

The PURGE command can be used to delete unused blocks and layers. The following is for information only—do not use on this drawing file.

Command: PURGE
Enter type of unused objects to purge
[Blocks/Dimstyles/LAyers/LTypes/Plotstyles/SHapes/textSTyles/Mlinestyles/All]: b
Enter name(s) to purge <*>:
Verify each name to be purged? [Yes/No] <Y>:

You can also use the Delete button in the Layer dialogue window to delete unused layers.

Arranging Chairs with ARRAY Command

You will use the ARRAY command to arrange chairs around the conference room. Using the illustration for guidance, do the following (Fig. 7-18 and Fig. 7-19):

1. ZOOM with a window to the conference room.

2 Use the CIRCLE command to draw a 6' diameter conference table.

3. INSERT a chair as illustrated (Fig. 7-18).

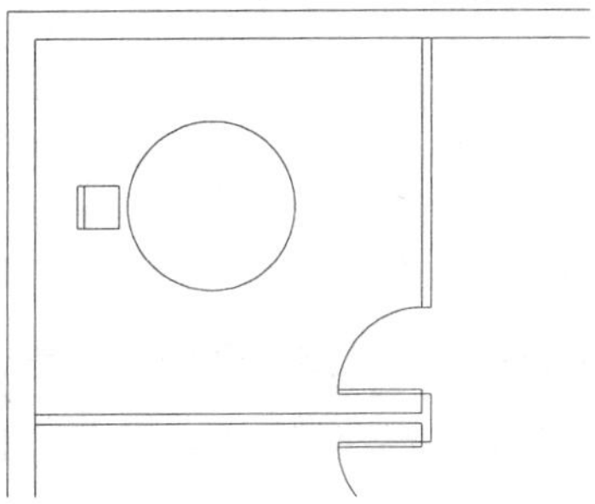

Fig. 7-18 Table and chair.

Command: **ARRAY**
Select objects: **(select chair)**
Select objects: **(press ENTER)**
Enter the type of array [Rectangular/Polar] <R>: **P**
Specify center point of array: **cen**
of **(point on edge of conference table or circle)**
Enter the number of items in the array: 10
Specify the angle to fill (+=ccw, -=cw) <360>: **(press ENTER)**
Rotate arrayed objects? [Yes/No] <Y>: **(press ENTER)**

Your drawing should appear as illustrated (Fig. 7-19).

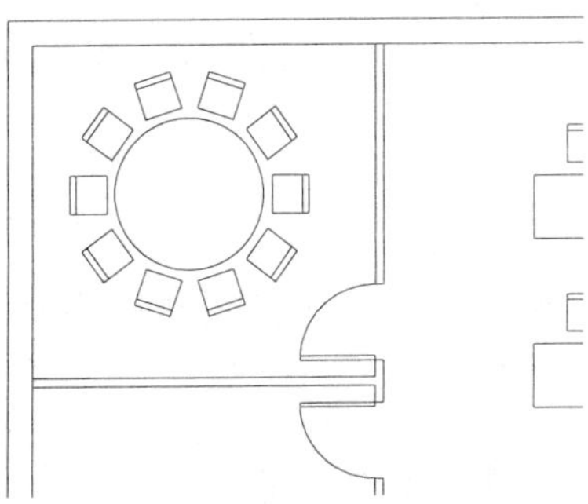

Fig.7-19 ARRAY of chairs.

Blocks with Attributes

Blocks can have attributes or text which gives additional information. These attributes can be constant or variable. Constant means that the same information is stored for each one while variable means that different information is stored depending on user input. Drawing symbol libraries of furniture companies use either constant or variable attributes. When you insert the block with variable attributes, you are prompted to provide the finish or other information. The information from these blocks then could be extracted to make specification lists. Facility managers use blocks with variable attributes to keep track of furniture or equipment inventory. You will use both types.

1. **Use ZOOM command to draw sofa in reception area (7-20).**
2. **Set SNAP to 3" and turn on.**
3. **Use the LINE command  to draw a two seater sofa 72" wide by 36" deep approximately as shown (Fig.7- 20).**

Tip: use object snap "midpoint" to draw line separating the cushions on sofa.

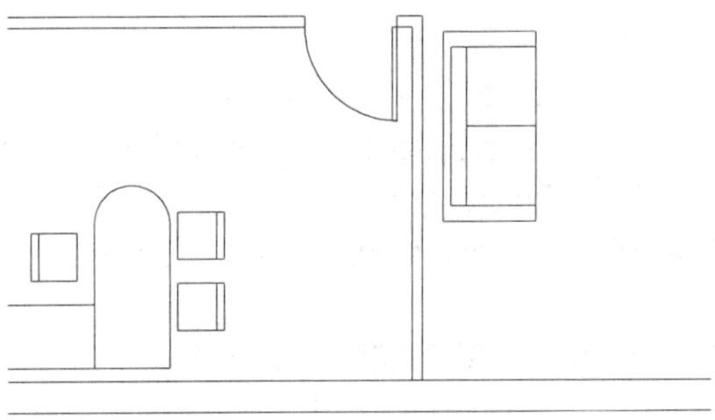

Fig. 7-20 Sofa.

ATTDEF Command

You will add the item number, manufacturer, and price as attributes of the sofa (Fig. 7-21). First you will change the value or mode of the attribute to a constant since every two seater sofa will have the same attribute tag information. After you enter the first attribute, you will repeat the command for the other items.

1. **Menu Bar > Draw > Block... > Define Attributes.**
2. **When dialogue window appears, type "Item#" in "Tag" box.**
3. **Type "F-1" in "Value" box.**
4. **Under Mode area, click on the box next to "Constant"until x appears.**
5. **Type "4" for Text Height.**
6. **Click on "Pick Point<" and the response will be:**

Start point: **(point A)**

7. Click on OK" to exit Attribute Definition window.

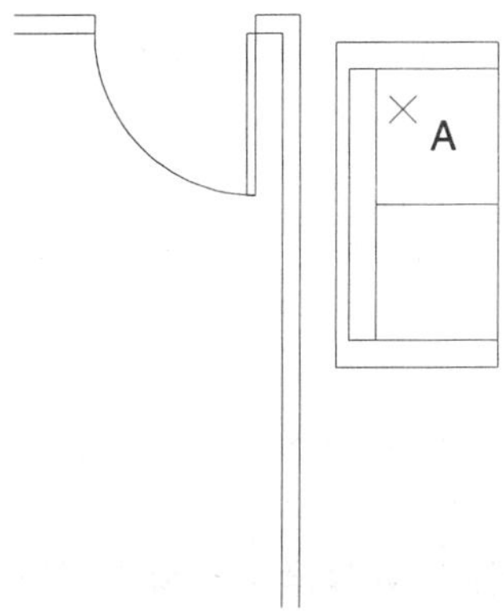

Fig. 7-21 Start point for attribute text.

You have created a "constant" attribute—meaning it is the same for each block. Now you will enter the manufacturer information:

1. Menu Bar > Draw > Block... > Define Attributes.
2. When dialogue window appears, type "MFGR" in "Tag" box
3. Type "CLAYTON" in "Value" box.
4. Click on "Pick Point<", the response will be:

Start point: **(click directly below ITEM# text)**

5. Click on "OK" to exit dialogue window.

Notice that you did not have to enter the height or constant—AutoCAD assumes what you entered before as the default. Next you will define the price attribute:

1. Menu Bar > Draw > Block... > Define Attributes.
2. When dialogue window appears, type "PRICE" in "Tag" box.
3. Type "1886" in "Value" box.
4. Click on "Pick Point<", the response will be:

Start point: **(click directly below MFGR text)**

5. Click on "OK" to exit dialogue window.

Your sofa with attribute text should appear as illustrated (Fig. 7-22)

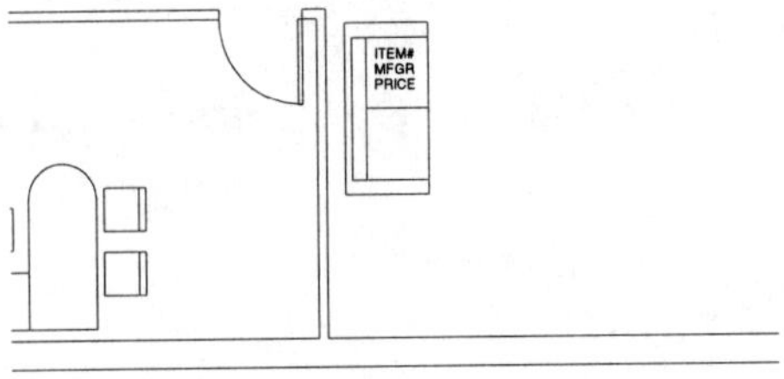

Fig. 7-22 Attributes.

Command: **WBLOCK**

1. Type "sofa" in Name box.
2. Click "Pick point" using lower left corner for insertion point.
3. Click "Select Objects" button.
4. Put a window around sofa and text and press Enter key or right click to end selection.
5. Click "Ok" to exit Block Definition dialogue window.
6. Erase the sofa block (because you will insert new ones).

Then insert the two sofa blocks:

1. Insert two "sofa" blocks as illustrated (Fig. 7-23).
2. Use the CIRCLE command to draw a table for the corner.

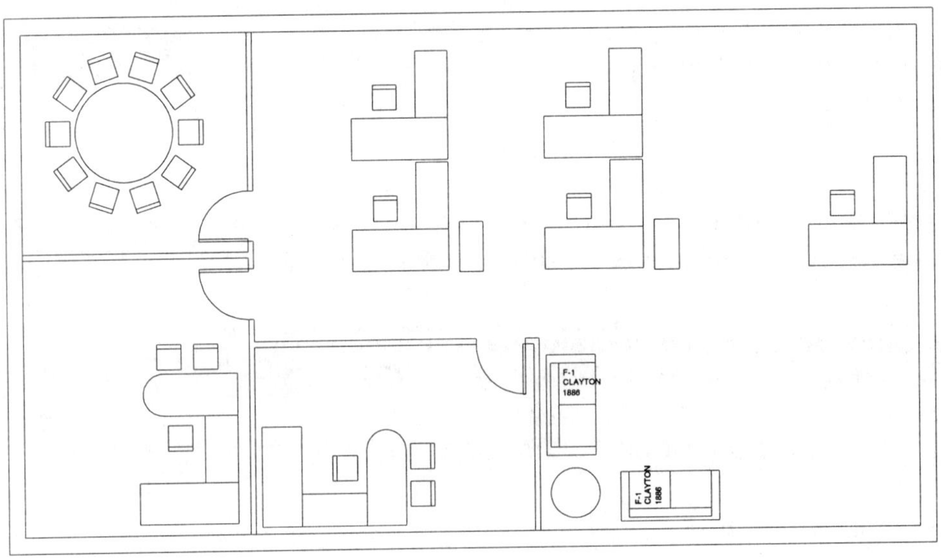

Fig. 7-23 Inserted sofa blocks with attributes.

In the next sequence, you will create a new layer “A-EQPM” to use to insert a computer block that has a variable attribute of a user name so you can name the person who has the computer:

Command: **LA**
LAYER

1. **Click on “New.”**
2. **In the layer list, type “A-EQPM” over the “Layer3” name.**
3. **Click on the color box.**
4. **Under the “Standard Colors”, click on the blue color.**
5. **Click on “OK” to exit the color window.**
6. **Check that the white box turned to blue for the A-EQPM layer.**
7. **While A-EQPM is still highlighted or selected, click on “Current.”**
8. **Click on “OK” to exit the Layer Properties Manager window.**
9. **Check that A-EQPM is the current layer.**

Next, you will make a simple box to indicate a computer:

1. **Change SNAP to 2"**
2. **With LINE Command, draw a rectangle 24" wide x 30" deep.**

Next, you will make a “variable” attribute for the computer block:

1. **Menu Bar > Draw > Block... >Define Attributes.**
2. **When dialogue window appears, type “user” in “Tag” box.**
3. **The “Constant” box in the mode section should be left blank. If an x appears, click on it to make it disappear.**
4. **Type “Enter computer user” in “prompt” box.**
5. **Click on “Pick Point<“, the response will be:**

Start point: **(point A on Fig. 7-24)**

6. **Click on OK” to exit dialogue window.**

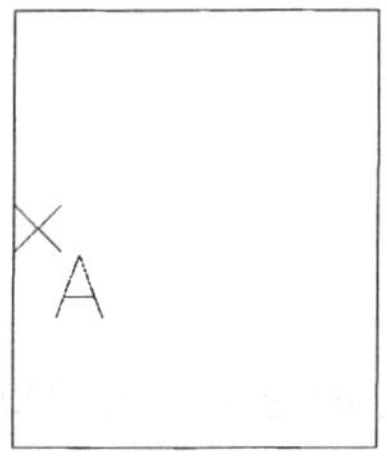

Fig. 7-24 Starting point for attribute text.

User should appear on the computer. Next, you will store the "user" block.

Command: **WBLOCK**

1. Type "user" in Name box.
2. Click "Pick point" button and use lower left corner for insertion point.
3. Click "Select Objects" button.
4. Put a window around the 4 lines and text and press Enter key or right click to end selection.
5. Click "Ok" to exit Block Definition dialogue window.
6. Erase the user block.

You have named the WBLOCK the same as the attribute tag. However, these could have been different. Now insert this block on each desk in the open office area (Fig. 7-25):

Command: **INSERT**

1. Scroll down list of names and select "user".

Specify insertion point or [Scale/X/Y/Z/Rotate/PScale/PX/PY/PZ/PRotate]: **(point to place on desk)**
Enter attribute values
Enter computer user: **SMITH**

1. Continue inserting the user block until all four desks have computers each with a different name.

Your office should appear similar to the illustration (Fig. 7-25).

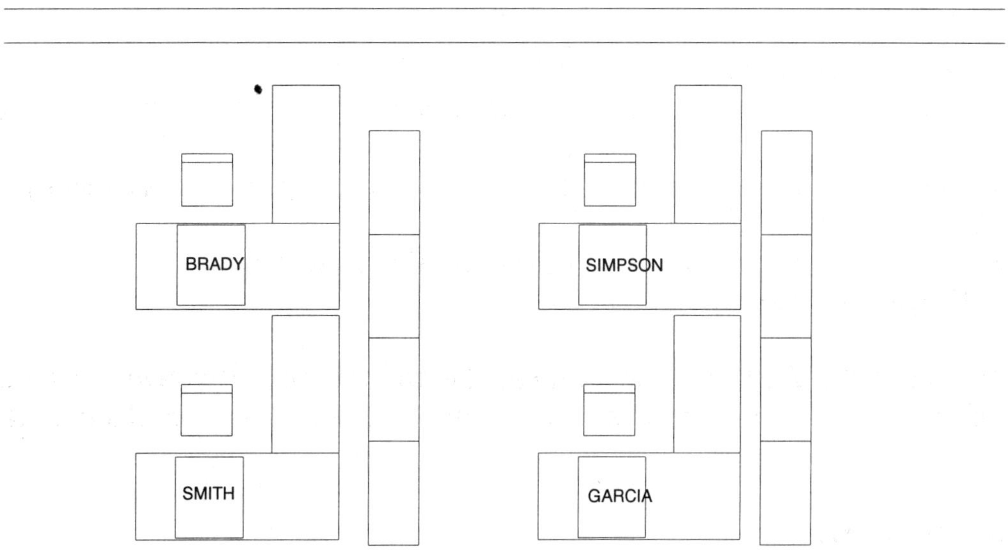

Fig. 7-25 Computers with user names.

ATTDISP Command

If you did not want to display the person's name or the furniture specifications on your drawing, you could use the ATTDISP command which will turn off the attribute text display.

> Command: **ATTDISP**
> Normal/ON/OFF <Normal>: **OFF**

The drawing regenerates without the user names or furniture information. Turn the display back on:

> Command: **ATTDISP**
> Normal/ON/OFF <Off>: **ON**

ATTEDIT, DDATTE Commands

ATTEDIT allows you to edit the attribute values that have variable (not constant) modes. You will change the name of the user from BRADY to SCOTT. The ATTEDIT command can be entered at the command prompt.

Command: **DDATTE**
Select block: **(Click on block with "BRADY" name)**

1. Type "SCOTT" in prompt box in "Edit Attributes" dialogue window.

Your drawing should appear as illustrated (Fig. 7-26).

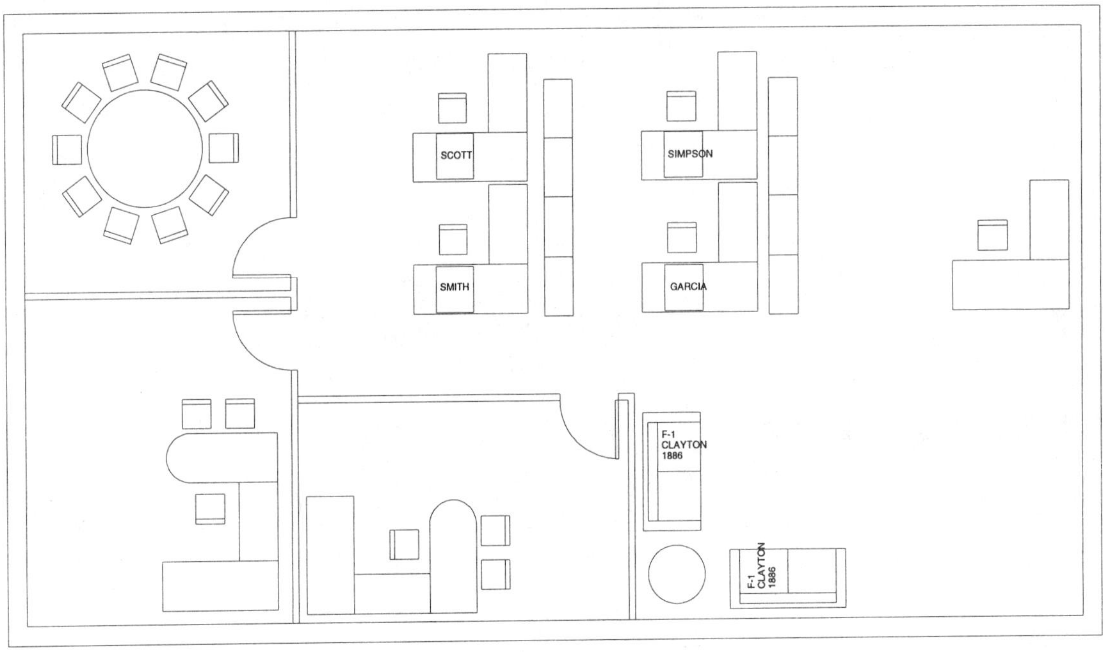

Fig. 7-26 Office with blocks and attributes.

XREF Command

The XREF command allows you to attach an external drawing to your drawing file as a reference. You can work with this external drawing without inserting it permanently into your drawing file. You are limited to scaling, rotating, and positioning the external reference drawing. So, it is similar to a block but has several advantages.

First, the external reference will not add the size of the file (in bytes) to your drawing. If you are working on a 200K drawing and attach a 400K external reference drawing or XREF, the size of your file will still be 200K, making saving operations faster. Had you added it as a block, the file would be 600K. For example, it might be used when you have a detailed logo that is large in file size. You can use an XREF for this logo rather than inserting it as a block into your drawing.

The second advantage of external references is that you can keep up with changes in the externally referenced drawing. That way several people can work on a project at once. For instance, the architect may be working on a building shell (Fig.7-27). The engineer would want to reference the building shell drawing while working on the electrical plan. As the interior designer working on the furniture plan, you might use the XREF command to attach the electrical plan drawing. Because the electrical plan already has the building shell as an external reference, you would have both the drawings as external drawing files (in a nested reference). Each time you reloaded your drawing, these external drawing files would be updated.

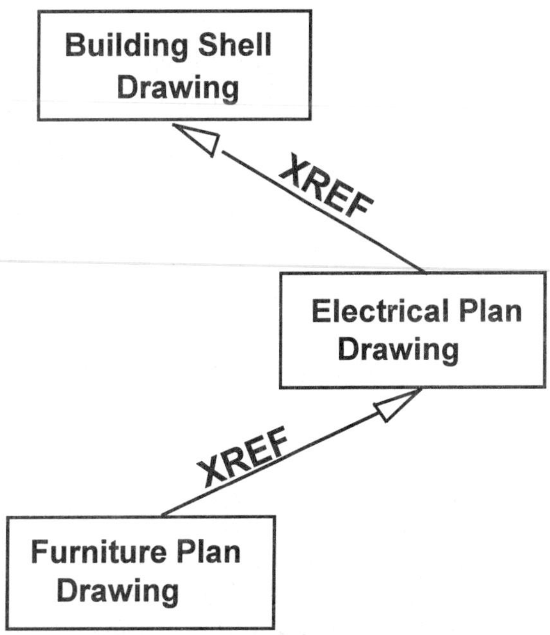

Fig. 7-27 Diagram of XREF.

The following is a procedure for attaching the rug created in Tutorial 4 to the furniture plan you created in this exercise. This makes some sense since that file occupies 720K of disk space because of all the hatching. The office plan is under 100K so attaching this file would save on disk space. The procedure below uses the menu bar although you could just type XREF at the command line to bring up the "External Reference" dialogue box.

1. Make A-FURN the current layer.
2. Menu Bar: Insert > External Reference...
3.Open the rug design created in Tutorial 4.
4. Click on "Browse" button, find the rug design file, click on "Open."
5. Click on "OK" to load the drawing file as an external reference.
6. Insert and rotate as needed (Fig. 7-28).

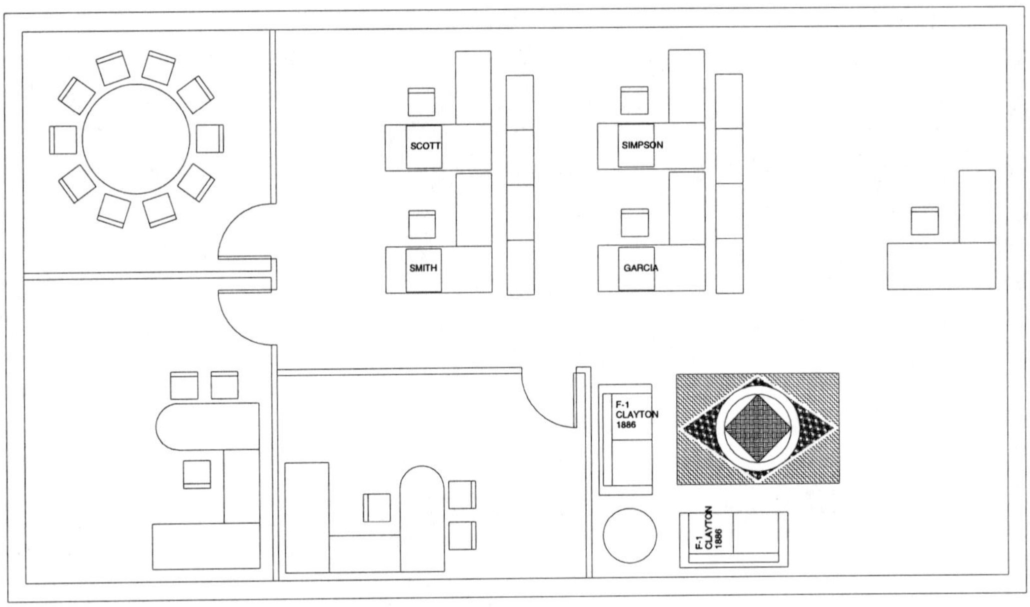

Fig 7-28 Final plan with rug design.

You are not able to erase parts of the rug because it is an external reference. However, you can scale, rotate, or move the rug. It is not a part of the drawing file, it is simply externally referenced. If you moved the xref or rug design file from its file location, AutoCAD will not be able to find it when it reloads your file and will just enter text of the location. If you want to keep it in its new location, you will need to detach the xref and reattach it.

This is the end of the tutorial. You will print the file and then detach the xref:

1. Print the file created with this tutorial (using Fit for scale).
2. Menu Bar: Insert > Xref Manager...
3. In the dialogue box, highlight the rug design file.
4. Click on "Detach" and check that file was cleared from list.
5. Click on "OK" to exit the "Xref Manager" window.
6. Save the file for the next tutorial.

TUTORIAL 8

DRAFTING A LIGHTING PLAN

Commands Learned: DDMODIFY
STRETCH
MATCHPROP
AREA
CAL
LINETYPE
LTSCALE
LWEIGHT

In this tutorial, you will learn more about layering a plan in order to create a lighting plan. You will draw with different linetypes and assign linetypes to layers. Blocks are affected by layers so you will learn more about them.

1. Open the drawing file used in Tutorial 7.

Command: **LA**

The "Layer Properties Manager" window appears.

1. Click on "New" button 5 times to create layers.
2. Click on the "Show details..." button.
3. Name the following layers and assign colors in the Detail area:

Layer Name	Assign Color
A-DOOR	**red**
A-WALL	**cyan**
A-CEILING-GRID	**yellow**
E-LITE	**yellow**
E-LITE-SWCH	**yellow**

5. Make A-CEILING-GRID the current drawing layer.

The furniture layer is displayed, but you do not need to see it so you will freeze this layer . Notice that the first two columns next to the layer name list contain two icons, light bulb and sun (Fig. 8-1).

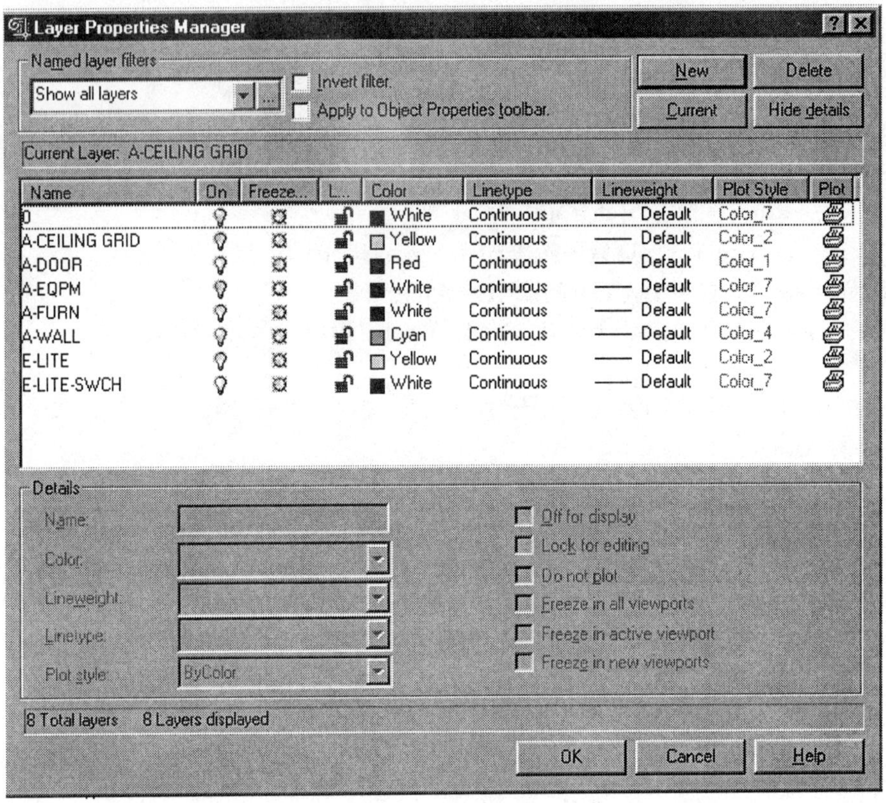

Fig. 8-1 Layer & Linetype Properties dialogue box .

The light bulb under the "On" heading signifies whether the display is on or off. The sun under the "Freeze..." signified that the layers are thawed (not frozen). When you click on the sun, a snowflake appears to indicate that the layer is frozen. Both have the same result in either displaying or not displaying the layer, but the "Freeze" option has an advantage in that AutoCAD will not use that layer in any calculations. This is the preferred option to use particularly when working in 3D.

1. Click on the sun icons to freeze the A-DOOR, A-FURN, and A-EQPM layers.
2. Click on "OK" to exit the dialogue box.

The furniture disappears but it is not lost, just the display of that layer. The display affects printing. If you printed now, you would get what you see on the screen or the walls and doors. But notice that the doors did not turn off even though you froze that layer; the reason is that you put them on the O layer which is not frozen. You will go back to the layer dialogue box to thaw that layer and then use a new command to change the doors to the doors layer.

DDMODIFY Command

This command, introduced earlier, allows you to change properties or other characteristics of any object.

Command: **LA**

1. Click on the snowflake icon to thaw the A-DOOR layer.
2. Click on "OK" to exit the dialogue box.
3. On the "Standard" toolbar click on the icon or type:

You will use DDMODIFY to change the doors to another layer. The icon is located on the Standard toolbar.

Command: **DDMODIFY**

The Properties window appears with "no selection" indicated on the top line. This can be docked as a viewport.

1. Select each door (a blue box or grip will appear on each door).
2. Click on the "Layer" line, click the down arrow and choose A-DOOR from the list.

The doors should turn to red indicating they are on this layer. Each line in the properties window is a property which can be changed. Some CAD users like to leave the Properties window docked next to the drawing area if they are doing a lot of editing. But you will close it:

1. Click on the X to close the Properties window.

STRETCH Command

The STRETCH command is useful to move portions of a drawing such as the door and door openings keeping endpoints and intersections intact. The trick in performing a successful STRETCH is to use the "window crossing" selection (or making a window from upper right to lower left) and to keep ORTHO and SNAP on. You will use stretch to move the door in the vice president's office to the other side. In the previous exercise you stored a view of that office so you will restore it:

Command: **VIEW**

1. **Click on the VP view.**
2. **Click on "Set Current" and "OK" buttons to exit View window.**
3. **Make sure ORTHO is on.**

Follow points in illustration (Fig. 8-2).

Command: **STRETCH**
Select objects to stretch by crossing window or -polygon. . .
Select objects: **(make a window with first point at A and second point at B)**
Select objects: **(press ENTER)**
Base point: **(point C)**
New point: **(point D)**

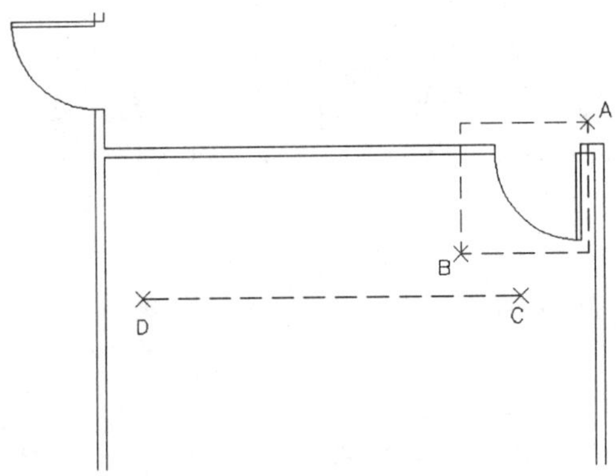

Fig. 8-2 Points for use with STRETCH command.

You will undo this command to return the door to its original location:

Command: **U**

Blocks, Colors, and Layers

Blocks take on the characteristics of the layer that they were created on. Suppose you had created a block named "arrow" to use for drafting arrows on the furniture layer but wanted to use it on another layer called "lighting" yet you wanted the furniture layer turned off. Then the block "arrow" would not show up because it was assigned to the turned off furniture layer. A block created on the "0" layer, though, has special characteristics. When inserted onto a layer, it is assigned to that layer and takes on its characteristics such as color and linetype. Consequently, remember to create blocks and written blocks on the "0" layer so that you can insert the block on any layer.

You assigned different colors for layers which is a good way to keep track of them. But you still could change color independent of your layers as you did in Tutorial 5 just by entering a color to the COLOR command. If you wanted to have a block with differerent colors in it, you would draw it on the 0 layer and change the colors with the COLOR command. When you inserted this block on a layer, it would keep the colors rather than displaying the color of the layer.

MATCHPROP Command

The MATCHPROP command allows you to match a property of one object with another. You will use it to match the layer property of the threshold to the walls:

1. Use EXPLODE command on exterior wall to revert to lines.
2. Draw an entry and threshold on the A-WALL layer similar to the illustration (Fig. 8-4).

Now click on the icon on Standard toolbar or type MATCHPROP as below:

Command: **MATCHPROP**
Select source object: **(click on any part of the threshold)**
Current active settings = Color Layer Ltype Ltscale Lineweight Thickness PlotStyle Text Dim Hatch
Select destination object(s) or [Settings]: **(select all the walls)**
Select destination object(s) or [Settings]: **(press ENTER or right click)**

The walls should now be the cyan color indicating that they are on the A-WALL layer. AutoCAD will change all the properties of the destination object. If you only want to change one or two, type in "S" to bring up the Settings dialogue box to select what you want matched. You will learn about some of these properties later in this book.

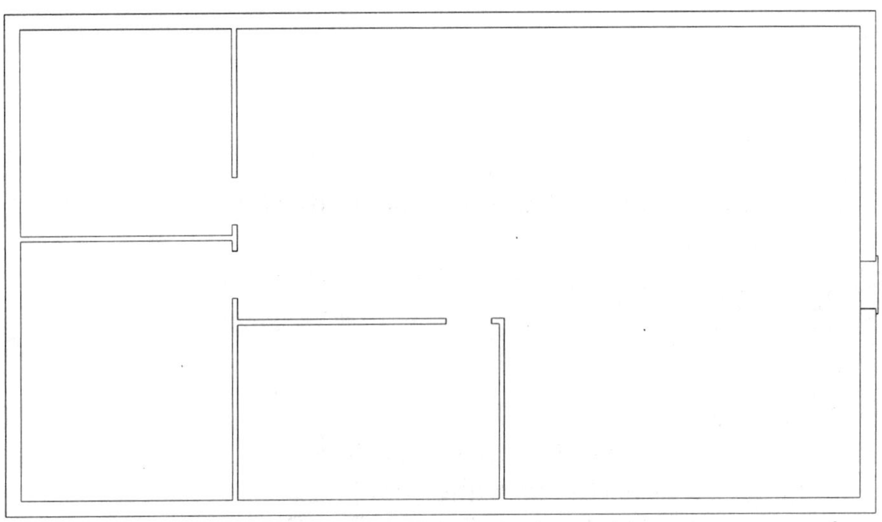

Fig. 8-3 Draw door opening and outside threshold.

Drafting A Lighting Plan

You will be drafting a quick lighting plan. To do so, you will draw the grid over the whole plan--not normally done with regular construction where the CAD designer would center the grid in each room and trim out the walls. However, in this case, you are using movable walls or partitions which are constructed after the ceiling grid is put in which is easier to draw.

1. **Zoom to drawing extents** **.**
2. **Check that A-CEILING-GRID is the current drawing layer.**
3. **Freeze the A-DOOR layer.**
4. **Set SNAP at 6" and turn on.**
5. **Turn OSNAP off on status line.**
6. **Check that ORTHO is on.**
7. **At 2' from lower wall, draft a horizontal line as illustrated (Fig. 8-4).**

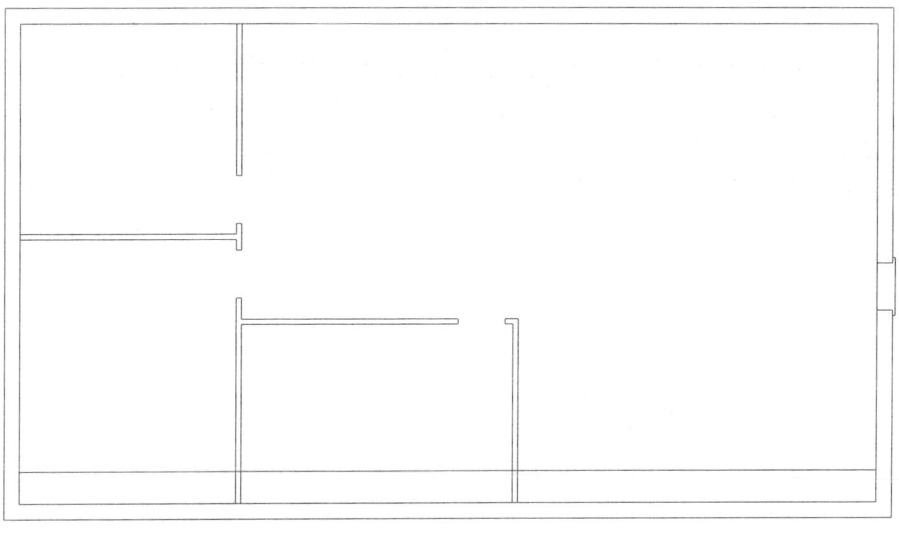

Fig. 8-4 Draw a horizontal line.

To speed up drafting, you will use the ARRAY command:

Command: **ARRAY**
Select objects: **L (or point to line you just drew)**
Select objects: **(press ENTER or right click)**
Enter the type of array [Rectangular/Polar] <R>: **R**
Enter the number of rows (---) <1>: **14**
Enter the number of columns (|||) <1> **(press ENTER)**
Enter the distance between rows or specify unit cell (---): **2'**

Then make the vertical lines and modify the grid:

1. Draft a vertical line 2' from left wall.

Then repeat the ARRAY command again:

Command: **ARRAY**
Select objects: **L (or point to vertical line you just drew)**
Select objects: **(press ENTER or right click)**
Enter the type of array [Rectangular/Polar] <R>: **R**
Enter the number of rows (---) <1>: **(press ENTER)**
Enter the number of columns (|||) <1> **27**
Specify the distance between columns (|||): **2'**

Your grid should look as in the drawing (Fig. 8-5).

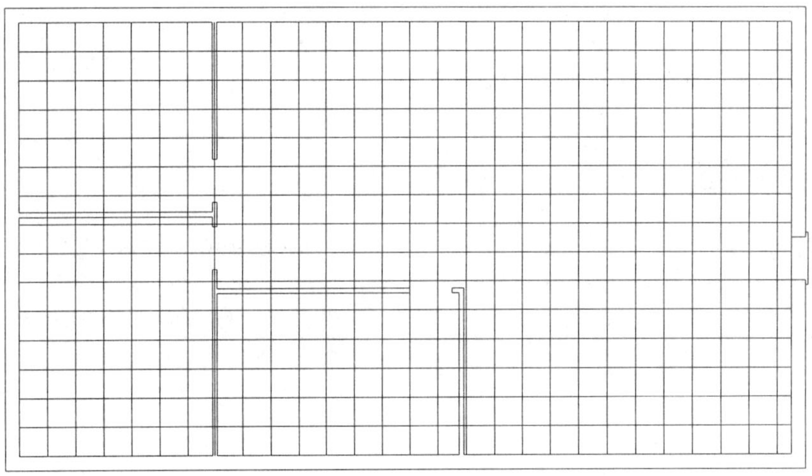

Fig. 8-5 Arrayed lines for 2x2 ceiling.

Next, you will lay in the lighting fixtures which are 2 x 2 fluorescent troffers. You are finished with the 2x2 grid so you can lock that layer to facilitate drawing. Locking the layer means that you cannot select the objects on that layer for editing so you will not select the grid lines by mistake.

1. On the Object Properties Toolbar, click on the 0 layer to make it the current layer.

Command: **LA**

1. Click on the unlocked icon to lock the A-CEILING-GRID layer.
2. Change SNAP to 2".
3. Draft a fluorescent fixture one like the one in the illustration, drawing the troffer about 2" inside the grid (Fig. 8-6).
4. Use the BLOCK command to make a block of it and name it 2x2 using the intersection of the grid as the insertion base point.
5. Make the "E-LITE" the current drawing layer.

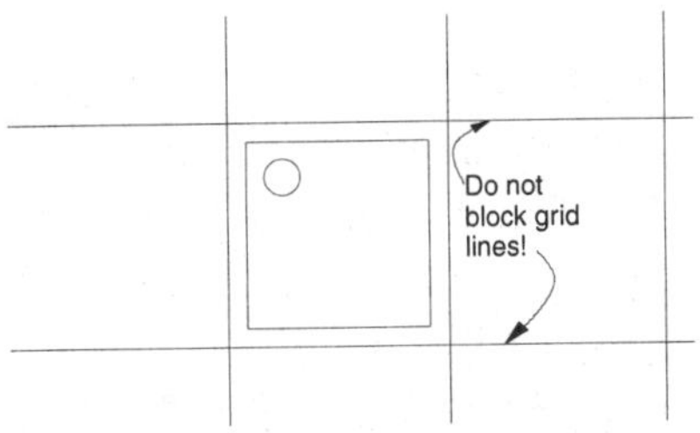

Fig. 8-6 Make a block of only the ceiling fixture.

Command: **LA**

1. Click on the lock icon to unlock the A-CEILING-GRID layer.

AREA Command

The AREA command is very useful for figuring space allocations for space planning. You tell AutoCAD whether you want to specify points around a space. Or, you can use the Object option to calculate areas for circles, rectangles (done with RECTANG command), polygons or other singular objects.

To calculate how many light fixtures you will need, you will calculate the area of the office. Use the four inside corners of the area covered by the grid. To make it easier, you will change OSNAP to "int" mode:

1. Zoom to drawing extents.
2. Turn SNAP on.
3. Menu Bar: Tools > Drafting Settings.
5. Click on "Object Snap" tab.
6. Check "intersection" and click on "OK" to exit dialogue box.
7. If you have not done so, use the EXPLODE command on the outside walls (otherwise, object snap will not work).

Command: **AREA**
Specify first corner point or [Object/Add/Subtract]:
Specify next corner point or press ENTER for total: **(point to inside corner of**

exterior wall)

Specify next corner point or press ENTER for total: **(point to next inside corner)**
Specify next corner point or press ENTER for total: **(point to next inside corner)**
Specify next corner point or press ENTER for total: **(point to last corner)**
Specify next corner point or press ENTER for total:**(press ENTER or right click)**

AutoCAD responds with:

Area = 237600.00 square in. (1650.0000 square ft.), Perimeter = 170'-0"

The "Add/Subtract" option of the AREA command is also useful. For instance, if you had to add up the area of a number of rooms, you would respond "Add" to the prompt. Then, each area would be measured and added to a total (or system variable "Area"). AutoCAD lists both the area for each room and then the running total area. The "Subtract" option will subtract any subsequent specified areas.

You will divide the area (in square feet) by 35 in order to get the number of fixtures that you will need to get 70 footcandles. This was derived from a lighting fixture catalog table. However, assume you do not have a calculator handy. AutoCAD will calculate it for you with CAL command.

CAL Command

The CAL command allows you to calculate although the calculator located in Windows accessories may be easier to use for simple calculations. The CAL command uses similar expressions as a calculator (i.e. + for add, - for subtract, * for multiply, / for divide). In the next sequence you will change OSNAP back to none and use the CAL command to divide the 1650 square feet by 35.

1. Click on the  icon on the object snap toolbar.

2. Check "Clear all" and click on "OK" to exit the object snap dialogue box.

Command: **CAL**
Initializing...>>Expression: **1650/35**

AutoCAD responds with:
47.1429

Thus, you will need about 47 fixtures in your office. You can do more sophisticated geometric calculations with the CAL command so if you are interested in learning more, refer to the AutoCAD User's Guide.

1. Change SNAP to 6" (alternate way is to use intersection object snap).

2. Use the MOVE command to move the block "2x2" in the grid in the lower left corner (Fig. 8-7). Or, you could insert the block.

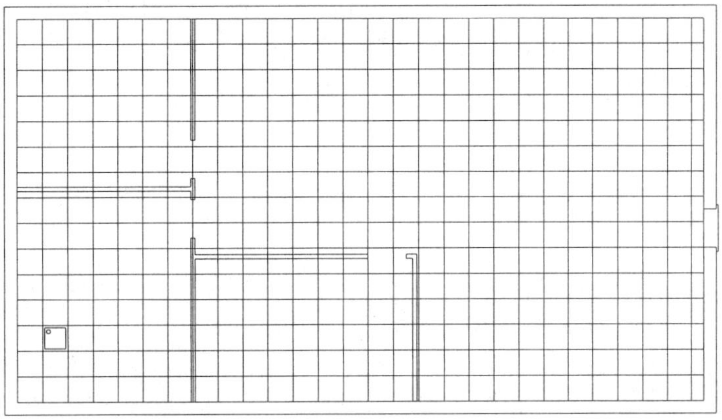

Fig. 8-7 Insert the fixture in the lower left corner of the grid.

Use the ARRAY command to lay out fixtures tentatively:

Command: **ARRAY**
Select objects: **L (or point to 2x2)**
Select objects: **(press ENTER or right click)**
Rectangular or Polar Array (R/P) <R>: **(press ENTER or right click)**
Number of rows (—) <1>: **6**
Number of columns (|||) <1>: **7**
Unit cell or distance between rows (—): **4'**
Distance between columns (|||): **8'**

Your fixtures should be inserted similar to the illustration (Fig.8-8).

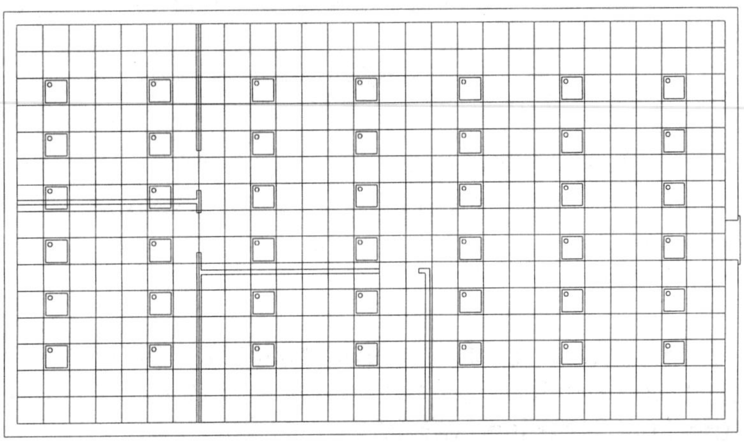

Fig. 8-8 Array of fixtures.

Of course, if you did not want to use the ARRAY command, another method would be to use the COPY command with multiple option. You will edit the layout of fixtures later. Next, you will draft the switching but instead of laboriously drafting the dashed lines, you will learn about assigning a linetype to a layer.

LINETYPE, LTSCALE Commands

AutoCAD automatically assigns linetype "continuous" to a layer. If you want to draw other types of lines such as a dashed, hidden, or center line, you must first load that linetype into your drawing file. Next, you need to change the scale of the linetype using the LTSCALE command because the scale is set at 1 so most lines will look continuous. Finally, you will create a new layer and assign the linetype to that layer. Obviously, you will only draw the assigned linetype for that layer because all lines will be drafted in the assigned linetype.

To draft the switching, you will need to load the "Dashed" linetype first and then change the scale of the linetype:

Command: **LAYER**

1. **Make E-LITE-SWCH the current drawing layer.**
2. **Click on "Continuous" on the E-LITE-SWCH layer.**
3. **Click on "Load..." button in "Select Linetype" window.**
4. **Scroll down and select "DASHED" by highlighting.**
5. **Click on "OK" to exit "Load or Reload Linetypes" window.**
6. **Highlight "DASHED."**
7. **Click on "OK" to exit "Select Linetype" window.**

"Dashed" should appear on the E-LITE-SWCH layer line under the linetype column. Now you will finish the switching plan quickly:

1. **Click on "OK" to exit "Layer Properties Manager" window.**

Next, you will change the linetype scale:

Command: **LTSCALE**
New scale factor <1.0>: **10**

1. **Zoom in to an area as illustrated (Fig. 8-9).**
2. **Turn ORTHO off.**
3. **Set SNAP to 2" and turn on.**
4. **Using the ARC command, draft a horizontal switching line connecting the two fixtures in the lower left corner (Fig. 8-9).**

5. Zoom to drawing extents.

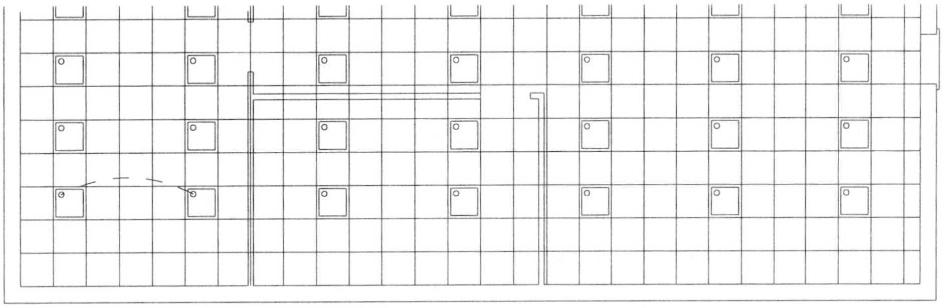

Fig. 8-9 Draw a switching line with ARC command.

You will use the ARRAY command to repeat the switching lines:

Command: **ARRAY**
Select objects: **L (or point to arc you just drew)**
Select objects: **(press ENTER or right click)**
Enter the type of array [Rectangular/Polar] <R>: **R**
Enter the number of rows (---) <1>: **6**
Enter the number of columns (|||) <1> **6**
Enter the distance between rows or specify unit cell (---): **4'**
Specify the distance between columns (|||): **8'**

Your array of switching lines will appear as in the illustration (Fig. 8-10).

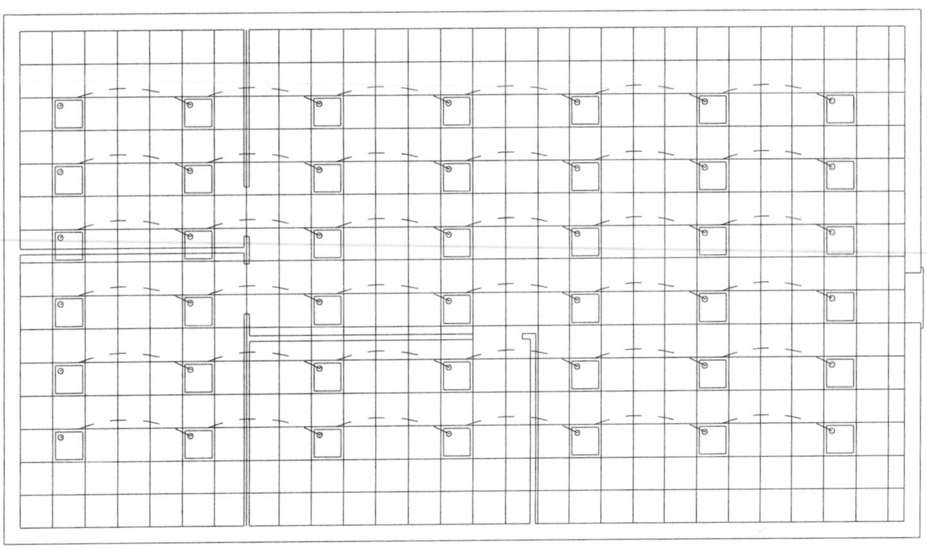

Fig. 8-10 Array of switching lines.

Now you will edit the fixtures and switching lines so they appear as illustrated (Fig. 8-11).

1. Use the ERASE command to remove some switching and fixtures (you may need to turn SNAP off).
2. Use the MOVE command to arrange fixtures better in office and conference room (keeping ORTHO and SNAP on).
3. Add switching using the ARC commnd to connect the rows to door locations.
4. Draw switches with DTEXT and LINE commands.

Your final lighting should appear as below (Fig. 8-11).

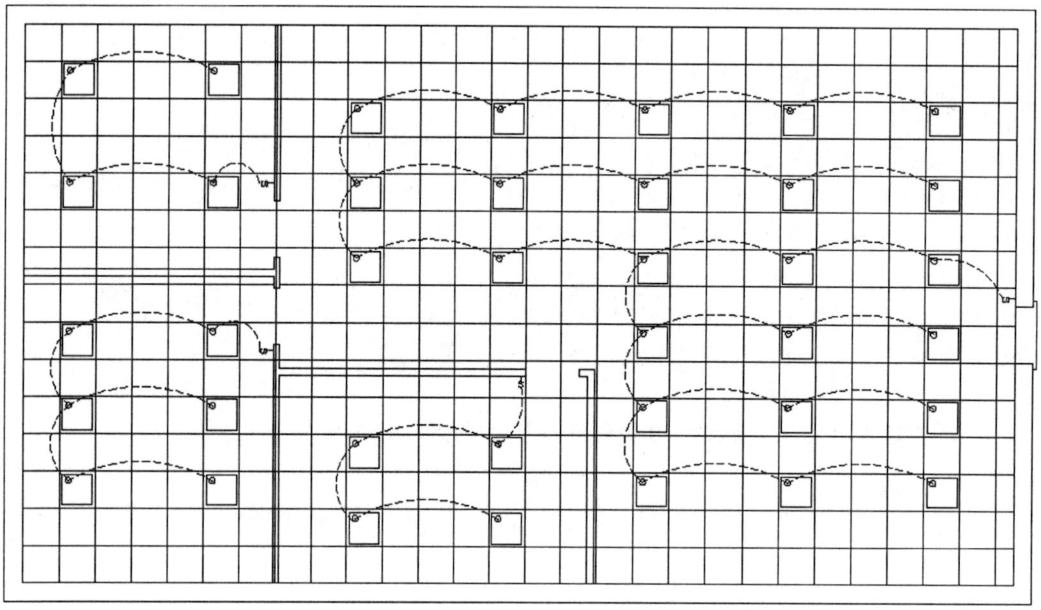

Fig. 8-11 Final lighting plan.

LINETYPE Command - Set Option

If you are drafting a number of lines of the same linetype, making a new layer and assigning the linetype to that layer makes sense. In the case where you just want to draw one or two lines with a different linetype, you can set a current linetype instead of assigning by layer. By setting a current linetype, all subsequent drawn entities will have that linetype.

1. Thaw the A-FURN layer and make it the current drawing layer.
2. Freeze the A-CEILING-GRID, E-LITE, and E-LITE-SWCH layers.

Use the set option to change the linetype from continuous to dashed:

Command: **LINETYPE**

1. Click on "Load" button.
2. Scroll down and select "dashdot."
3. Click on "Current" to make dashdot the current linetype.
4. Click on "OK" to exit the "Layer & Linetype Properties" dialogue box.
5. Use the LINE command to draw a line.

The line appears dashed even though continuous is assigned to the "furniture" layer.

1. Use the ERASE  **command to erase the line.**

Be sure to return the linetype to the bylayer option or else all your lines will be dashdot from now on!

Command: **LINETYPE**

1. Select "ByLayer" at the top of the list.
2. Click on "Current" to make "ByLayer" current.
3. Click on "OK" to exit the "Linetype Manager" dialogue box.

Now you must once again assign linetypes by layer. If you need to use different linetypes on a block, use this option to draft them on the "0" layer which always has linetype "continuous" assigned to it.

LWEIGHT Command

The LWEIGHT command makes different lineweights. It is best to assign these by layer because some of the lineweights are too thin to see on the screen so you might be using a thick line without knowing it if you just set the lineweight. You will assign a .7 thickness to the A-WALL layer but you cannot see the thick lineweight until you click on the LWT button on the status line.

1. Use LAYER command and highlight A-WALL layer.
2. In Lineweight column, click on "____Default."
3. In Lineweight window, highlight ".70 mm" and click "OK" to exit.
4. Click on "LWT" button on status line to turn on.

The walls should now appear to have thick lines although your walls may appear a lot thicker than the illustration (Fig. 8-12). The appearance on your monitor is not tthe same as what is plotted which will be the .7 mm. thickness. Lineweights are best assigned right before plotting.

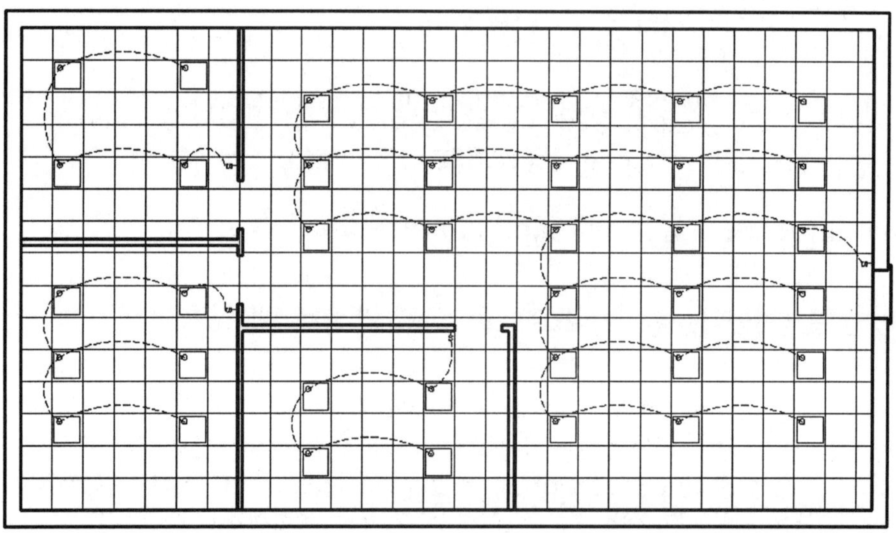

Fig. 8-12 Walls with heavy lineweight.

You will turn the walls back to the default option and print out your lighting plan:

1. Use LAYER and assign "Default" or single lines back to A-WALL layer.
2. Thaw the A-CEILING-GRID, E-LITE, E-LITE-SWCH layers.
3. Make A-CEILING-GRID layer the current drawing layer.
4. Freeze the A-FURN layer.
5. Print the drawing using Fit as the scale to print on letter-size or A-size paper.

TUTORIAL 9

DRAFTING ELEVATIONS AND ISOMETRIC DRAWINGS

Commands Learned:	UCS	POINT	ID
	XLINE	MEASURE	LIST
	PDMODE	DIVIDE	STATUS
	PDSIZE	RENAME	TIME

You will learn to draw on an angle, use construction lines to draft elevations, and use an isometric grid to draft a lateral file.

1. Open the drawing file you worked on in the previous tutorial.
2. Thaw the A-FURN layer.
3. Make a new layer called PLANTER and make it the current drawing layer.
4. Freeze the A-CEILING-GRID, E-LITE, and E-LITE-SWCH layers.
5. Turn the attribute display off (ATTDISP - off) if it is on.

Drawing on Angles

So far, you have drafted only in a horizontal or vertical position. To draw on an angle, you do not want to turn ORTHO off as it will result in jagged diagonal lines although it is possible to use SNAP to get straight and angled lines. There are several methods to draw on an angle similar to using a triangle in traditional drafting:

UCS - Z option (rotates the UCS).
SNAP - Style Option with Isometric style.

In the right corner of the office plan, you will be drawing a diagonal line on a 45 degree angle. This will be a corner for plants. In conventional drafting, you would have used your 45 degree triangle. You will be rotating the UCS or drawing plane.

UCS Command

You turned off the UCS icon so turn it back on again:

> Command: **UCSICON**
> Enter an option [ON/OFF/All/Noorigin/ORigin] <OFF>: **ON**

The UCS (User Coordinate System) command lets you determine the location of the drawing plane. By examining the UCS icon, you can see that the X coordinate direction is horizontal and the Y direction is vertical. What you cannot see is the Z direction which is going straight back toward you. By rotating the UCS or drawing plane around the Z axis, you can rotate the grid on an angle. In this exercise, you will only be rotating the X-Y drawing plane to draw on angles. However, you will learn more about the UCS when you draw in three dimensions. For now, rotate the drawing plane 45 degrees around the Z axis. You can use the UCS icon on the Standard Toolbar or type:

> Command: **UCS**
> Current ucs name: *NO NAME*
> Enter an option [New/Move/orthoGraphic/Prev/Restore/Save/Del/Apply/?/World] <World>: **Z**
> Specify rotation angle about Z axis <90>: **45**

Notice that the UCS icon rotates to show the angle that you are drawing on.

1. Make sure ORTHO is on.
2. Keep SNAP off.
3. Draft a diagonal line for start of planter overlapping wall line (Fig. 9-1).
4. Use OFFSET command  to offset diagonal line 3" (Fig. 9-1).

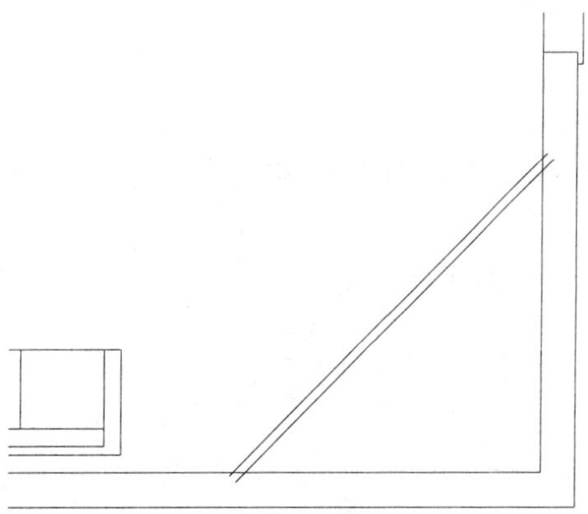

Fig. 9-1 Draw two diagonal lines.

If ORTHO was not on, these lines would be jagged. You will return the drawing plane back to the normal drawing plane or World Coordinate System by entering:

Command: **UCS**
Current ucs name: *NO NAME*
Enter an option [New/Move/orthoGraphic/Prev/Restore/Save/Del/Apply/?/World]:
(press ENTER)

The World Coordinate System is always the default user coordinate system in the UCS command so anytime you want to return to it, simply press the ENTER key to the UCS data prompt. In the next sequence, you will trim the planter lines. The plants were drawn with the CIRCLE command, using the LINE command starting at the center of the circle. One branch was drawn and arrayed with the polar option of the ARRAY command. Finally the plant was blocked, inserted, and scaled down. If you had saved your Exercise 5 and did a good plant, you could alternatively just go back and WBLOCK that plant to use in this exercise.

1. Use TRIM command to make sure planter lines connect to wall and do not overlap.
2. Zoom in very close to make sure lines connect (no gaps).
3. Make the A-FURN the current drawing layer.
4. Using CIRCLE and LINE commands (not SKETCH), draw a plant (Fig. 9-2).
5. Make a block of the plant with the BLOCK command.
6. Use the INSERT command, check "Specify On-Screen" for scale.
7. Insert the plant block and use a scale of .75 for X and press ENTER for Y.
8. Insert the plant block again but use a different scale.

Your plan should look like the illustration (Fig. 9-2).

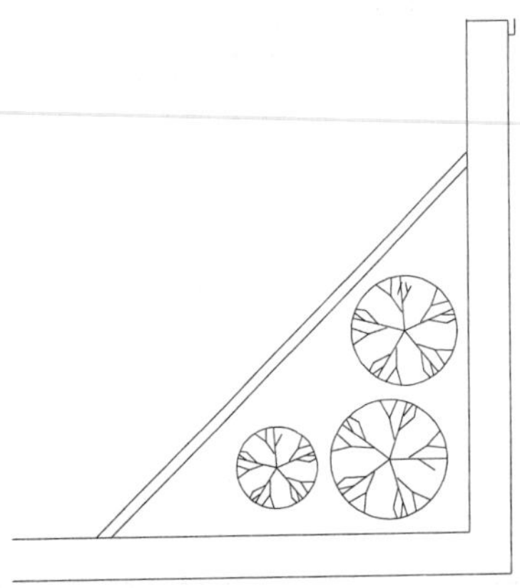

Fig. 9-2 Draw a plant and use smaller scale to make different sizes.

1. Make a new layer named A-GLAZ and make it the current drawing layer.
2. Make sure ORTHO is on and SNAP is on and set to 2" increment.
3. Draw a window using the LINE command.
4. Make a block of the window with the BLOCK command.
5. Use INSERT command to insert windows in reception and office areas (Fig. 9-3).

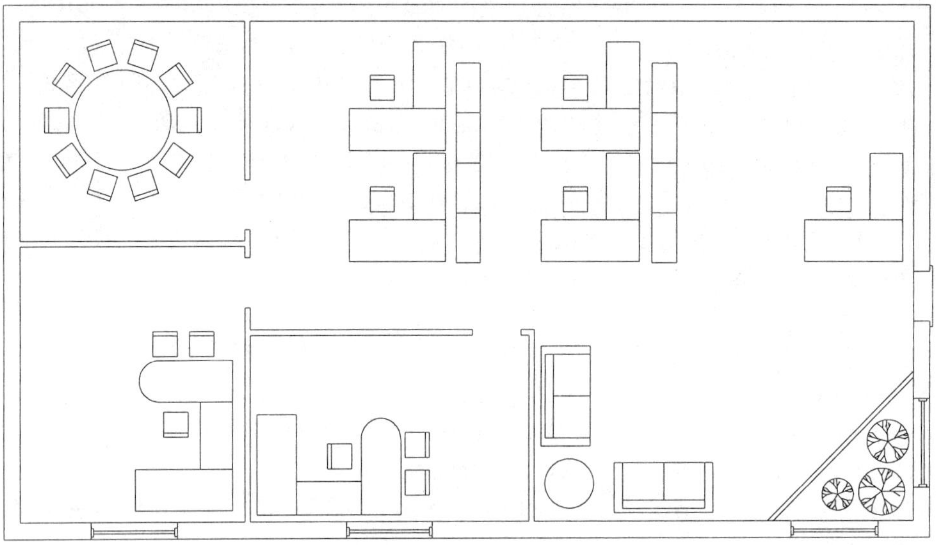

Fig. 9-3 Plan with windows.

You will change the size of the edit selection box or pickbox:

1. Menu Bar: Tools > Options... > Selection Tab.
2. Decrease the Pickbox Size by moving the bar to the left a couple of notches.

Next, you will draw a floor pattern with the BHATCH command. You will need to enclose the boundary around the room first:

1. **Make A-FURN the current drawing layer.**
2. **Zoom with a window to see just the conference room.**
3. **With ORTHO and SNAP on, use the LINE command to draw a line to close off the conference room so you can apply a hatch pattern. Make sure the line connects precisely to the walls by using "int" object snap (Fig. 9-4)**

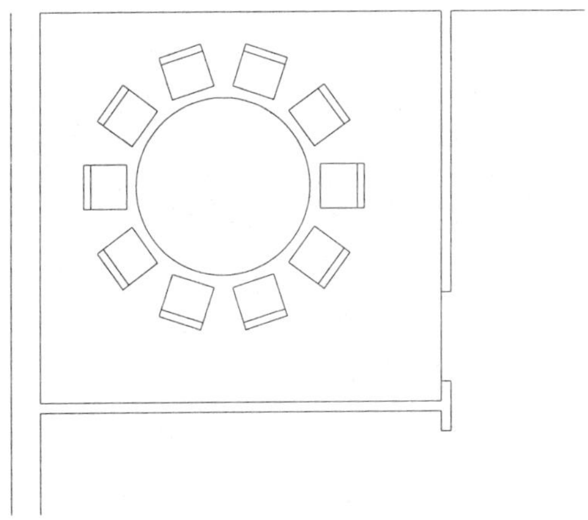

Fig. 9-4 Draw an enclosed boundary.

1. **Menu Bar: Draw > Hatch...**
2. **Click in box next to "Swatch" and click on "Other Pre-defined Tab."**
3. **Scroll down list and click on "net" pattern.**
4. **Click on the box next to "Scale:" and type in 96 (very important!).**
5. **Click on "Pick Points <" button.**

AutoCAD will respond with:

Select internal point: **(point in an emply space between furniture and walls)**

AutoCAD will respond with:

Select internal point: Selecting everything...
Selecting everything visible...
Analyzing the selected data...
Analyzing internal islands...
Select internal point: **(press ENTER)**

1. Click on "Preview" button.
2. Press ENTER key.
3. If hatch is correct, select "OK" button.

The result should look as in the illustration (Fig. 9-5).

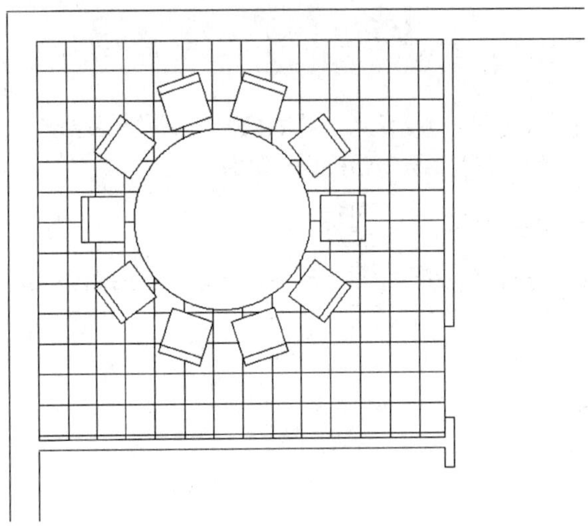

Fig. 9-5 Completed floor pattern in conference room.

In drafting, many areas to be hatched are complicated. You will want to experiment with the different styles of hatching such as normal, outer, and ignore to figure out the best way to hatch. The "Preview" option will save you time. If you are having major difficulties, you could make a new layer and make a polyline to use as an enclosed boundary for the hatch. As a last alternative, use the EXPLODE command to allow you to use edit commands such as TRIM, EXTEND to refine the hatch.

The hatch patterns are scaled to the plotted size of the drawings. You entered 96 for scale (1/8"=1' scale) which gave you a 12" tile floor. By entering 48 for scale, you would get a 6" tile floor. You typically would put a scale of 96 for plotting at 1/8"=1' scale (8 x 12), 48 for 1/4"=1' scale (4 x 12), 24 for 1/2"=1' scale, etc.

Associative hatching was checked when you did this hatch. Associative hatching means that when you move the furniture, the hatch will be automatically edited to reflect the new boundaries. You wil try this:

1. Use the MOVE command and make a window (lower left to upper right) around the table and chairs and then move them.

You should see the hatch redone to accommodate the new location. Keep this new location of the furniture if you like it or undo it.

Drafting Elevations

When you work in 2D, drafting elevations is similar to conventional drafting.

Command: **LA**

1. Make a new layer named A-FURN-ELEV and make it the current drawing layer.
3. Assign the color blue to the A-FURN-ELEV layer.
4. Turn SNAP (set at 2") and ORTHO on.
5. Zoom to drawing extents .
6. Pan the floor plan down so you can see the area to draw an elevation (Fig. 9-6).

Information for elevation:

Files: 52" high
Desks: 28" high
Ceilings: 9' high.
Conference Table: 30" high

In the next sequence, you will make construction lines that project off of the floor plan.

XLINE Command

The XLINE command draws different types of construction lines. Use the xline icon on the draw toolbar or type:

Command: **XLINE**
Specify a point or [Hor/Ver/Ang/Bisect/Offset]: **V**
Specify through point: **(line up crosshairs with inside wall and point)**
Specify through point: **(line up crosshairs with inside wall and point)**

You should have two construction lines similar to the illustration (Fig. 9-6).

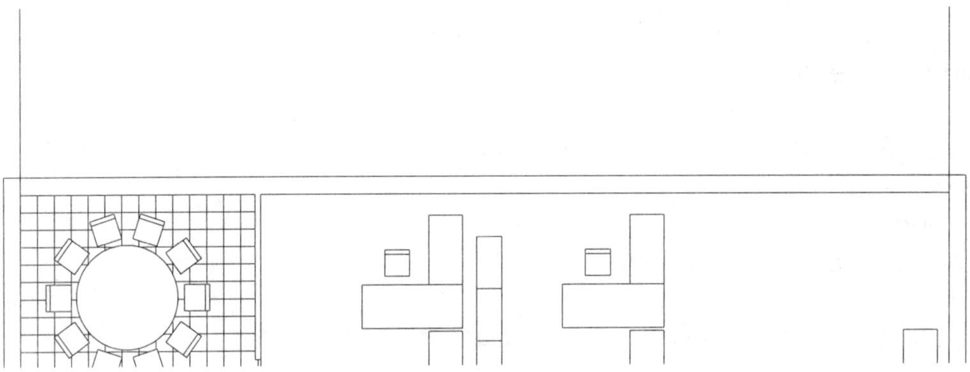

Fig. 9-6 Construction lines.

1. Draw an elevation with a 9' ceiling height (Fig. 9-7).
2. Erase the construction lines.

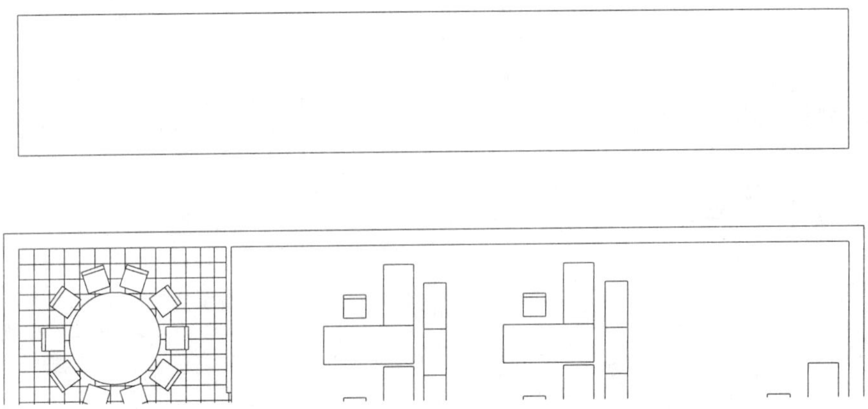

Fig. 9-7 Elevation outline.

Now you will use the XLINE command to draw the conference table (Fig. 9-8):

Command: **XLINE**
Specify a point or [Hor/Ver/Ang/Bisect/Offset]: **V**
Specify through point: **quad (or click on  on object snap toolbar)**
Specify through point: **(line up crosshairs with edge of conference table)**
Specify through point: **(line up crosshairs with other edge of conference table)**
Specify through point: **(press ENTER or right click)**

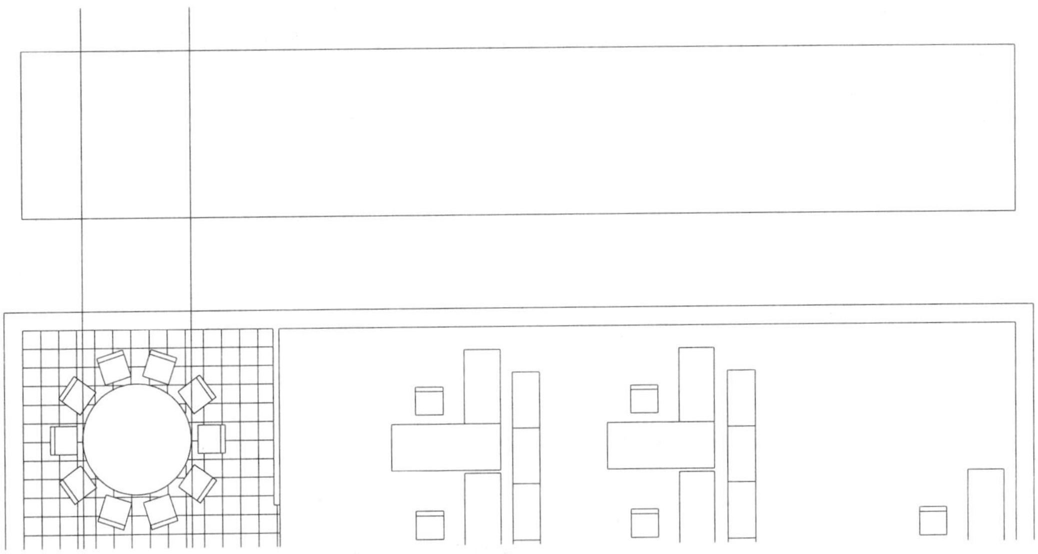

Fig. 9-8 Construction lines projected off conference table.

1. Draw the conference table top 2" thick by 30" high (Fig. 9-9).

Fig. 9-9 Conference table top.

1. Erase the construction lines.
2. With the aid of the XLINE command (or just line up the crosshair), finish the elevation similar to the illustration (Fig. 9-10) changing the value of SNAP as needed.
3. For the conference base and chair, draw one side and use the MIRROR command.
4. On the interior wall, use OFFSET after you draw the first line.
5. Use the BLOCK command to store the painting.

Your final elevation may look similar to what is illustrated (Fig. 9-10).

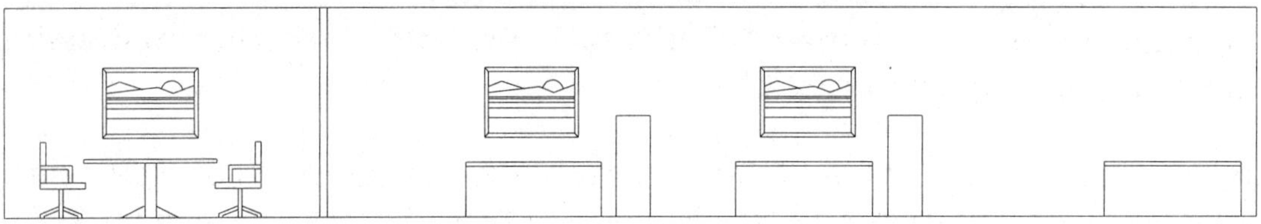

Fig. 9-10 Final Elevation.

1. Save your drawing file .

Drawing on Angles Again in Isometric

Do the following to get ready to draft one of the lateral files in isometric:

1. Turn ORTHO off .
2. Set SNAP at 1".

Command: **LA**

1. Make a new layer named A-FURN-ISO and make it the current drawing layer, assigning a color to it.

By using the "Style" option of the SNAP command, you can set up an isometric grid. This grid lets you draw only in a vertical and either the right or left direction on a 30 degree angle. You will first draw the left side of the file or the left and vertical direction.

Command: **SNAP**
Specify snap spacing or [ON/OFF/Aspect/Rotate/Style/Type] <0'-1">: **S**
Enter snap grid style [Standard/Isometric] <S>: **I**
Specify vertical spacing <0'-1">: **(press ENTER or right click)**

The grid should rotate to isometric.

1. ZOOM with a dynamic window to the right of the plan about the area illustrated (Fig. 9-11).

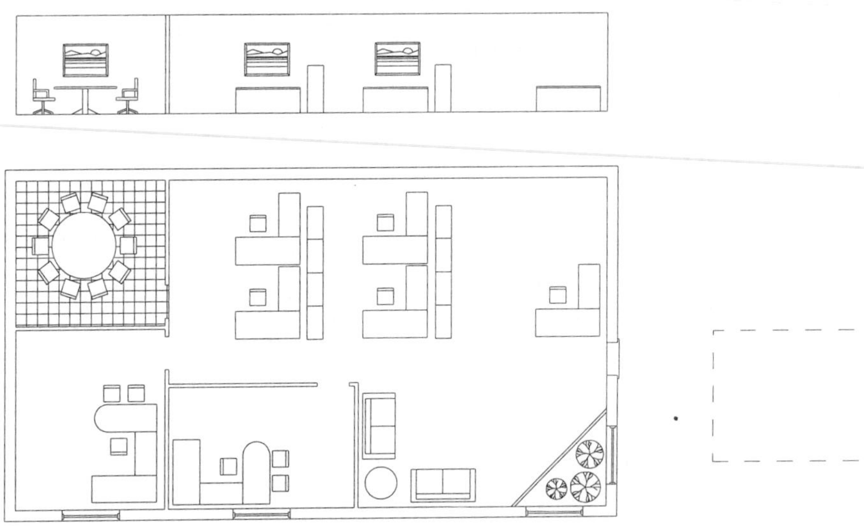

Fig. 9-11 Zoom to a small area to right of the plan.

Now, you will draft the file which is 36" wide by 18" deep by 52" high. You will be using the PLINE command (so you can offset the front of the file) and using polar coordinates:

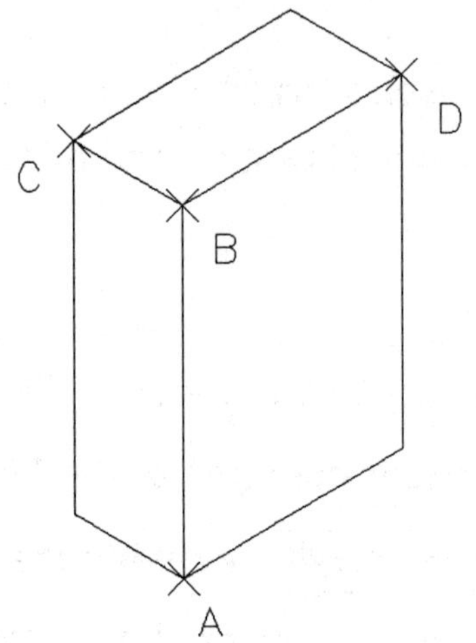

Fig. 9-12 Points for making the polyline.

Command: **PLINE**
Specify start point: **(point in lower area of screen to start)**
Current line-width is 0'-0"
Specify next point or [Arc/Close/Halfwidth/Length/Undo/Width]: **@52<90**
Specify next point or [Arc/Close/Halfwidth/Length/Undo/Width]: **@36<30**
Specify next point or [Arc/Close/Halfwidth/Length/Undo/Width]: **@52<-90**
Specify next point or [Arc/Close/Halfwidth/Length/Undo/Width]: **@36<210**
Specify next point or [Arc/Close/Halfwidth/Length/Undo/Width]: **(press ENTER)**
Command: **(press ENTER)**
PLINE
Specify start point: **INT (or use**  **icon)**
of **(Point A)**
Specify next point or [Arc/Close/Halfwidth/Length/Undo/Width]: **@18<150**
Specify next point or [Arc/Close/Halfwidth/Length/Undo/Width] **@52<90**
Specify next point or [Arc/Close/Halfwidth/Length/Undo/Width]: **INT (or use** 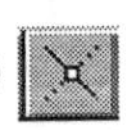 **icon)**
of **(Point B)**
Specify next point or [Arc/Close/Halfwidth/Length/Undo/Width]:: **(press ENTER)**

Command: **(press ENTER)**
PLINE
Specify next point or [Arc/Close/Halfwidth/Length/Undo/Width]: **INT (or use icon)**
of **(Point C)**
Specify next point or [Arc/Close/Halfwidth/Length/Undo/Width]: **@36<30**
Specify next point or [Arc/Close/Halfwidth/Length/Undo/Width]: **INT (or use icon)**
of **(point D)**

1. **Zoom with a window to see just the file.**
2. **Use OFFSET to offset the front of the file 1" inside (Fig. 9-13).**
3. **Use EXPLODE command to explode the inside polyline.**
4. **Use MOVE to move the bottom line 2 inches up (Fig. 9-14).**
5. **Use TRIM command to trim two lines at bottom.**

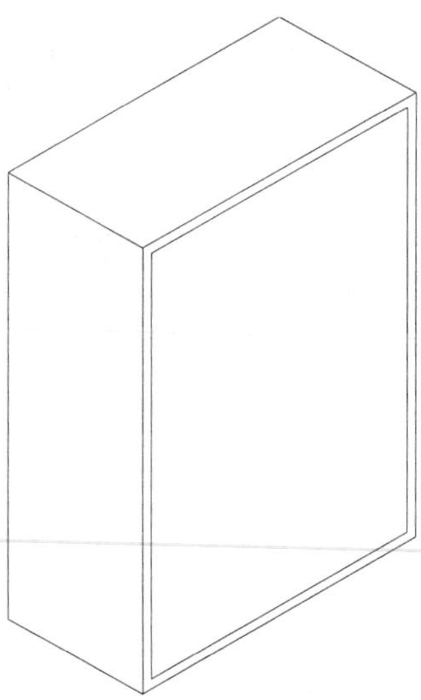

Fig. 9-13 Offset front of file.

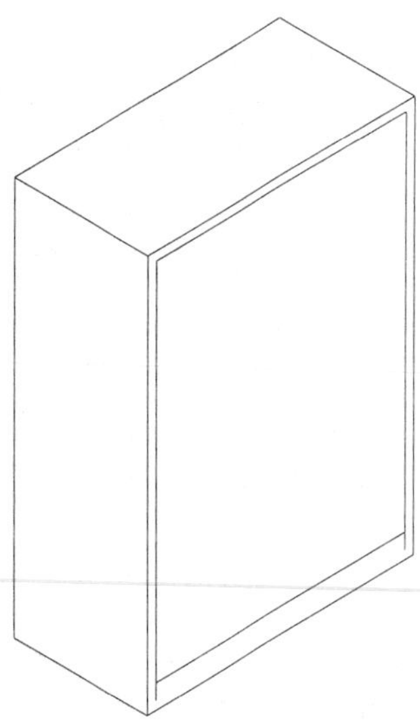

Fig. 9-14 Move bottom line 2 inches.

The four file drawers are 12" high with no space in between. To draw these, you will enlist the aid of the MEASURE command. In order to use it, you also need to learn about the POINT command and PDMODE system variable.

POINT, PDMODE, PDSIZE Commands

The POINT command draws points but the appearance of these points is governed by the PDMODE system variable or command. These commands are easiest to enter through the Point Style window.

1. **Menu bar: Format > Point Style..."**
2. **Click on top row, third icon from the left or the cross.**
3. **Check box next to "Set Size in Absolute Units."**
3. **Click on "OK" to exit.**

This icon menu allows you to change the size of the point side either in absolute units or setting the size relative to the screen. You set the size in absolute units so that when you zoom the whole drawing, your points will not be large. This is the same as changing the system variable PDSIZE. You will accept the default size. To see what a point looks like, you will use the POINT command. The icon is located on the Draw toolbar or type:

Command: **POINT**
Specify a point:**(point anywhere)**

Had you not changed the point style, this point would be very small. The point is useful combined with "node" object snap to draw objects. However, you will not use it so erase it:

Command: **E**
ERASE
Select objects: **L**

Now you will use the MEASURE command to measure points at 12" intervals.

MEASURE Command

This command measures an object placing points at specified intervals. Follow the point in the illustration which shows how the measured points will look when finished (Fig. 9-15).

1. **Menu Bar: Draw > Point... > Measure.**

MEASURE
Select object to measure: **(point A)**
Specify length of segment or [Block]: **12**

You should see markers placed every 12" along the line. The result should look as in the illustration (Fig. 9-16). You also can insert blocks by choosing the “block” option and then you would be prompted for a segment length to space the blocks apart.

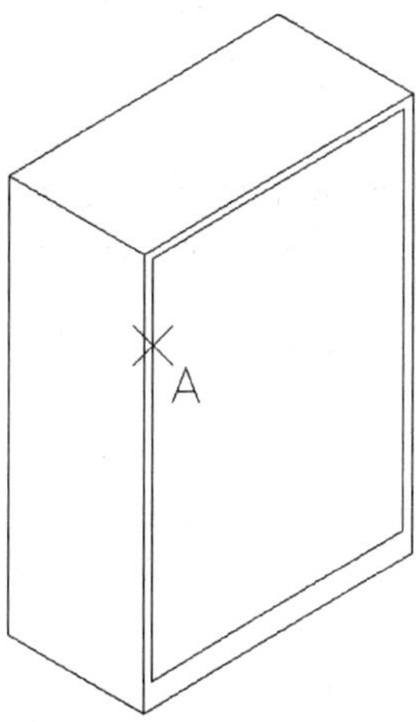

Fig. 9-15 Select this line.

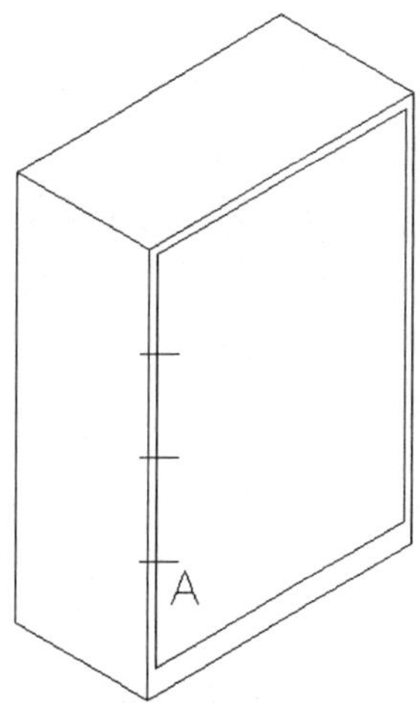

Fig. 9-16 Measured points.

DIVIDE Command

A closely related command is the DIVIDE command which divides an entity based on a number you enter. You will just try this out by drawing a circle which you will erase later.

1. Draw a circle anywhere on the screen.
2. Menu Bar: Draw > Point... > Divide.

Select object to divide: **(select circle)**
Enter the number of segments or [Block]: **6**

The block option allows you to insert a block instead of the point. For instance, you might have inserted a block made of two arcs to make a daisy. The DIVIDE command is thus useful to align objects such as lines, arc, text along an arc, line, circle, or curved line.

1. Erase the circle.

Drawing in Isometric

So far, you have not used the isometric grid crosshair but you will in the next sequence to draw the drawers (Fig. 9-17). The crosshairs only allow you to draw to the left or top so you will toggle the isoplane to the left with the F5 key.

1. **Makes sure ORTHO .**
2. **Menu Bar: Tools > Drafting Settings... > Object Snap Tab.**
3. **Check that "intersection" and "endpoint" are checked.**
4. **Check Object Sanp On (or press F3) to turn OSNAP on.**

Command: **L**
LINE
Specify first point: **(point at first marker shown in Fig. 9-17)**
Specify next point or [Undo]: **(press F5 key to toggle crosshair to right, click OSNAP off on status line and draw line to right extending over vertical line)**
to point: **(press ENTER)**

1. **Zoom in and use TRIM to make line connect precisely (Fig. 9-17).**

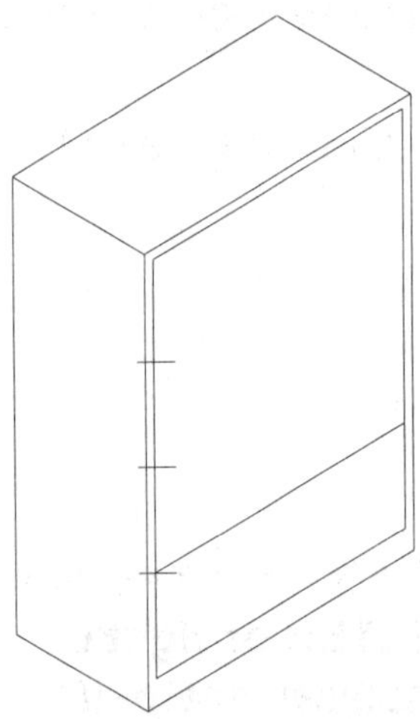

Fig. 9-17 Use isometric grid to draw the lines for the file drawer.

1. **Use the XLINE command to draw a construction line at "midpoint" of bottom line (Fig. 9-18).**
2. **Zoom with a window to draw pull.**
3. **Set SNAP at 1/2" and turn on.**

4. Use the LINE command to draw the pull approximately as shown (Fig. 9-18).

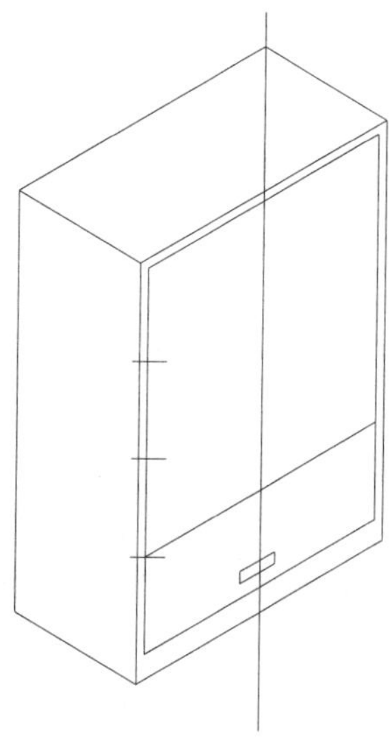

Fig. 9-18 Use construction line to help you draw the drawer pull.

1. Use ERASE command to erase the construction line.

You will finish the lateral file by using the COPY command and use your points as markers (Fig. 9-19).

1. Turn OSNAP on.

Command: **COPY**
Select objects: **(select bottom drawer pull and line above for drawer line)**
Select objects: **(press ENTER or right click)**
Specify base point or displacement, or [Multiple]: **M**
Specify second point of displacement or <use first point as displacement>: **(point to lower left corner of drawer)**
Specify second point of displacement or <use first point as displacement>:**(point to second marker)**
Specify second point of displacement or <use first point as displacement>:**(point to last marker)**
Specify second point of displacement or <use first point as displacement>:**(press ENTER or right click)**

1. Use ERASE command to erase the three markers or points.
2. Click on OSNAP on status line to turn off.

Your file should look similar to the illustration (Fig. 9-19).

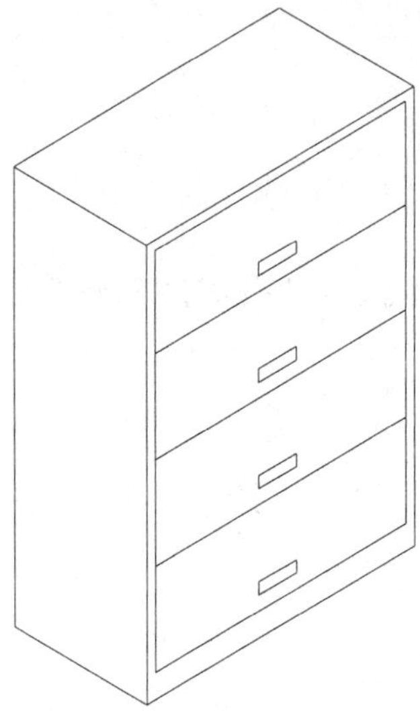

Fig. 9-19 Final drawing of lateral file.

Be sure to return the SNAP to the standard grid by the following command sequence:

Command: **SNAP**
Specify snap spacing or [ON/OFF/Rotate/Style/Type] <0'-0 1/2">: **S**
Enter snap grid style [Standard/Isometric] <I>: **S**
Specify snap spacing or [Aspect] <0'-0 1/2">: **(press ENTER or right click)**

1. Zoom to drawing extents **.**

The next section will discuss some inquiry commands that will provide useful information about the drawing and its contents.

LIST Command

The LIST command is used for examining the data that is stored for any entity. On the Standard Properties toolbar where the distance icon is found is the flyout for the list icon or you can just type at the command prompt:

Command: **LIST**
Select objects: **(point to a block)**

BLOCK REFERENCE Layer: "FURNITURE"
Space: Model space
Handle = 2A5
"DESK"
at point, X= 37'-3" Y= 24'-7" Z= 0'-0"
X scale factor 1.0000
Y scale factor 1.0000
rotation angle 0
Z scale factor 1.0000

This command is very useful if you do not know which layer an object is on or the name of a block. The information listed will be different for different entities. For a line, you would see starting and end points. For hatch, the hatch pattern, scale, etc. To save time, you can select more than one object. The information for the first selected object is listed first, then the second, and so forth.

1. Click on the X in the upper right corner to close the text window.

AutoCAD Text Window

The AutoCAD text window is also a very useful inquiry tool. You can scroll back to see all the commands used in a session. You can also copy to NotePad or a word processing program in case you had some serious problems with a drawing session and wanted to share with someone for analysis. Anytime you want to see the text window, press the F2 function key. You can then minimize or close the window by clicking in the upper right corner of the window.

STATUS Command

While the LIST command gives information about an entity, the STATUS command reports various defaults, modes and other useful information about your drawing. Below is the information given for a file similar to this one:

1. Menu Bar: Tools > Inquiry > Status...

An area to check is free disk space. AutoCAD sets up temporary files and needs extra disk space to swap to disk space while you work on your drawing. A minimum of 30 megabyes of free disk space should be allowed, but this is somewhat dependent on the complexity of the drawing.

1. Press ENTER to return to command prompt.
2. Click on the X in the upper right corner to close the text window.

RENAME Command

When naming a block or layer you may find that you would like to change its name. In the sequence below, the PLANTER layer was renamed to PLANT.

Command: **RENAME**

1. Click on "Layers..."
2. Highlight Planter and check that it appears in box next "old Name:"
3. Type in "Plant" in box next to "Rename To:"

The "0" layer cannot be renamed nor can the "continuous" linetype. Perhaps the easier way to rename a layer is in the "Layer Properties Manager" dialogue box:

Command: **LA**

1. Click on the layer PLANT until it is highlighted.
2. Click on "Show details" if details are not already displayed.
3. Click in the box next to "Name" and type in PLANTER to change the name again.

ID Command

The ID command will display the coordinate location of a point. Of course, you could have just put the crosshairs over the point and make a note of it but this is more precise.

1. Menu Bar: Tools > Inquiry > ID Point.

Command: '_id
Specify point: **(point to a location)**

TIME Command

Lastly, the TIME command allows you to see your elapsed drawing time, time when you first created the drawing, and elapsed time for the current drawing sesstion.

1. Menu Bar: Tools > Inquiry > Time.

Command: '_time
Current time: Monday, February 09, 1998 at 11:54:04:520 AM
Times for this drawing:
Created: Friday, February 02, 1996 at 11:42:21:630 AM
Last updated: Monday, April 22, 1996 at 11:37:04:710 AM
Total editing time: 1 days 04:33:29.430
Elapsed timer (on): 1 days 04:33:29.430
Next automatic save in: 0 days 00:10:44.860
Display/ON/OFF/Reset: **(press ENTER or right click)**

1. Click on the X in the upper right corner to close the text window.

This is the end of this tutorial.

1. Save this file for the next tutorial.
2. Print using "Scaled to Fit" for scale.

TUTORIAL 10

DIMENSIONING FLOOR PLANS AND SECTIONS

Commands Learned:	DIMLINEAR	DIMTEDIT
	DIMCONTINUE	DIMRADIUS
	QDIM	QLEADER

In this tutorial, you will use AutoCAD's associative dimensioning feature.

1. Open your drawing file from the previous tutorial.
2. Create a new layer named A-WALL-DIMS and make it the current drawing layer.
3. Zoom leaving about 4 or 5 feet around the floor plan.
4. Freeze all layers except 0, A-DOOR, A-GLAZ, A-FURN-ELEV, A-WALL, A-WALL-DIMS, and PLANTER.
5. Use the EXPLODE command  **to explode the outside walls created with multilines if you did not do that earlier.**

Just the walls, windows, and doors should be showing on the plan.

AutoCAD has five basic types of dimensioning: linear, angular, diameter, radius, and ordinate. You will primarily use linear dimensioning so this tutorial concentrates on this type of dimensioning.

Dimensioning Terms

It is important to know basic dimensioning terms because you can alter their appearance or values when you establish a dimension style (Fig.10-1). The plan (Fig. 10-2) shows what your dimensioned drawing should look like when finished.

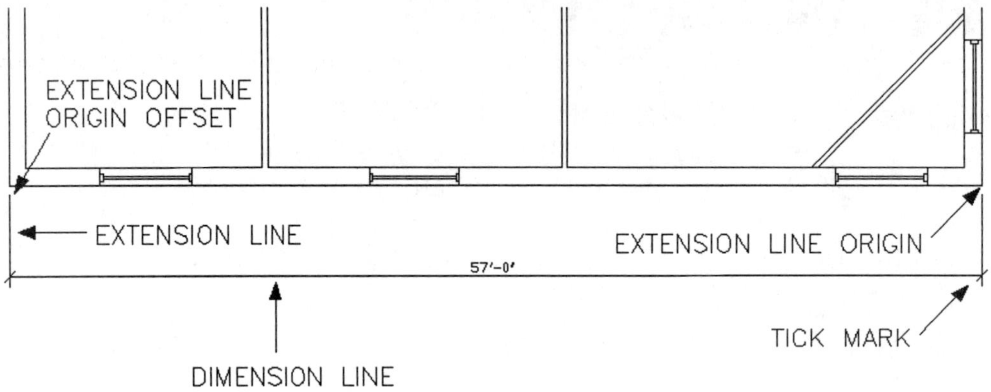

Fig.10-1 Dimensioning terms.

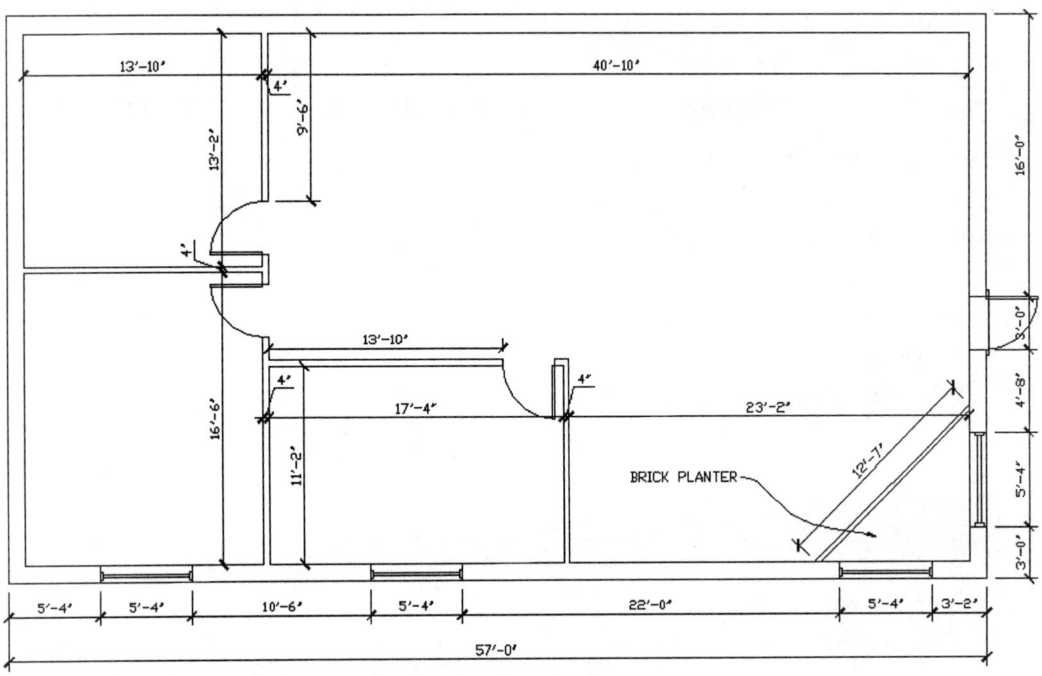

Fig. 10-2 Final dimensioned plan.

Creating a Dimension Style

1. Menu Bar: Format > Dimension Style...
2. In Dimension Style Manager window, click "New."
3. Type "PLAN" for "New Style Name."
4. Select "Standard" for "Start With."
5. Select "All dimensions" for "Use for."
6. Click "Continue" button to exit "Create New Dimension Style" window and Click Line and Arrows Tab (if not showing).

The new Dimension Style window appears:

7. Follow instructions on illustration (Fig. 10-3) to change:
Arrowheads: "Architectural tick"
Arrow size" 6"
Extend beyond dim lin: 6"
Offset from origin to 6"

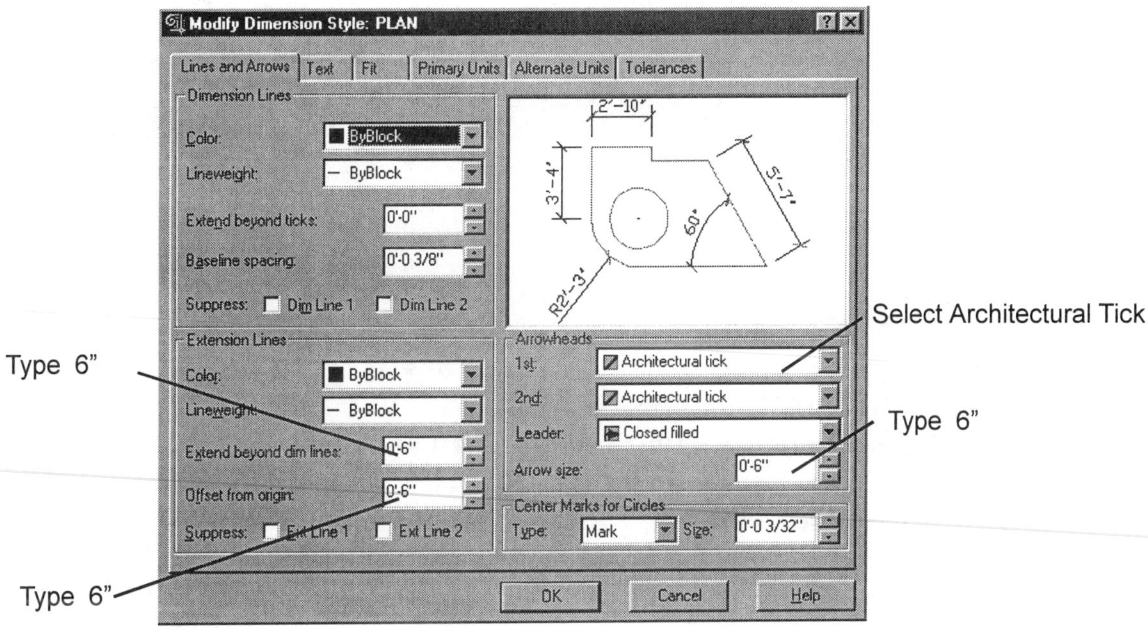

Fig. 10-3 Change settings of Lines and Arrows.

Using 6" will make the tick size about 1/8" when plotted at scale 1/4"=1'-0" scale. Likewise, your extension lines will extend 1/8" beyond the dimension lines and be offset 1/8" from the origin. Had you wanted to plot at 1/8"=1', you would use.values of 12" so tick size would be 1/8" when plotted.

8. Click on “Text” tab.

9. Follow instructions on illustration (Fig. 10-4) to change:

Text height: 6”

Vertical: Above

Offset from dim line: 3”

Text Alignment: Aligned with dimension line.

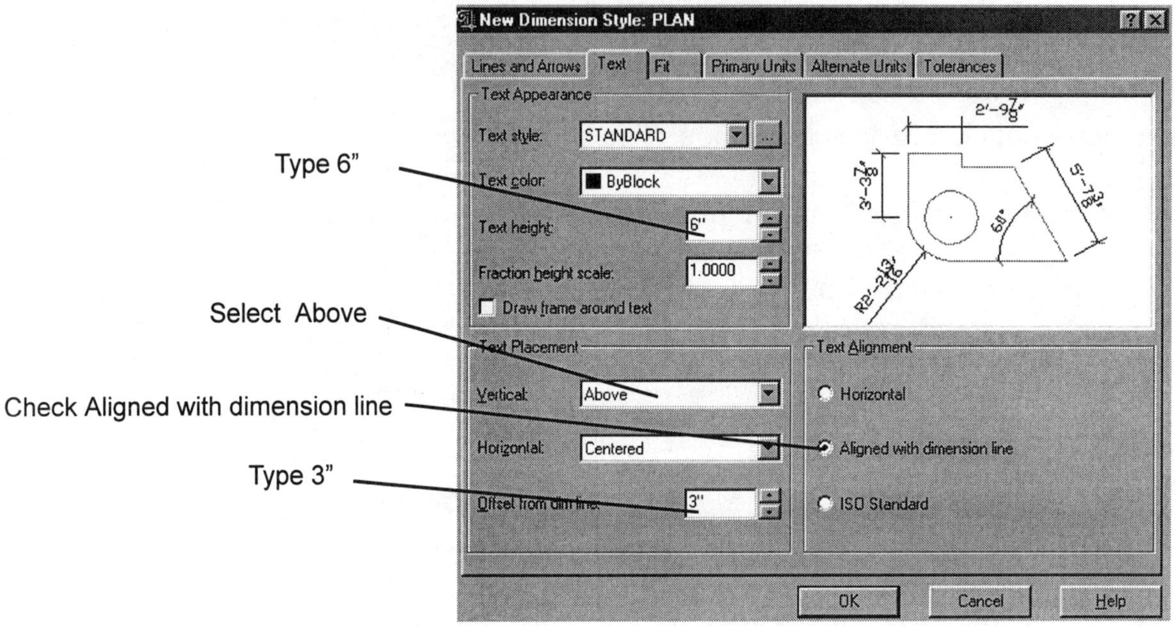

Fig. 10-4 Change settings of Text.

Again, the text height will be 1/8” high when you plot this drawing at 1/4”=1’-0” scale because you set text to 6” height.

10. Click on “Primary Units” Tab.

11. Follow Instructions on illustration (Fig. 10-5) to change values:

Primary Units: Architectural

Precision: 0’-0”

Round Off: 0’-0”

Remove check from Inches in Zero Suppression

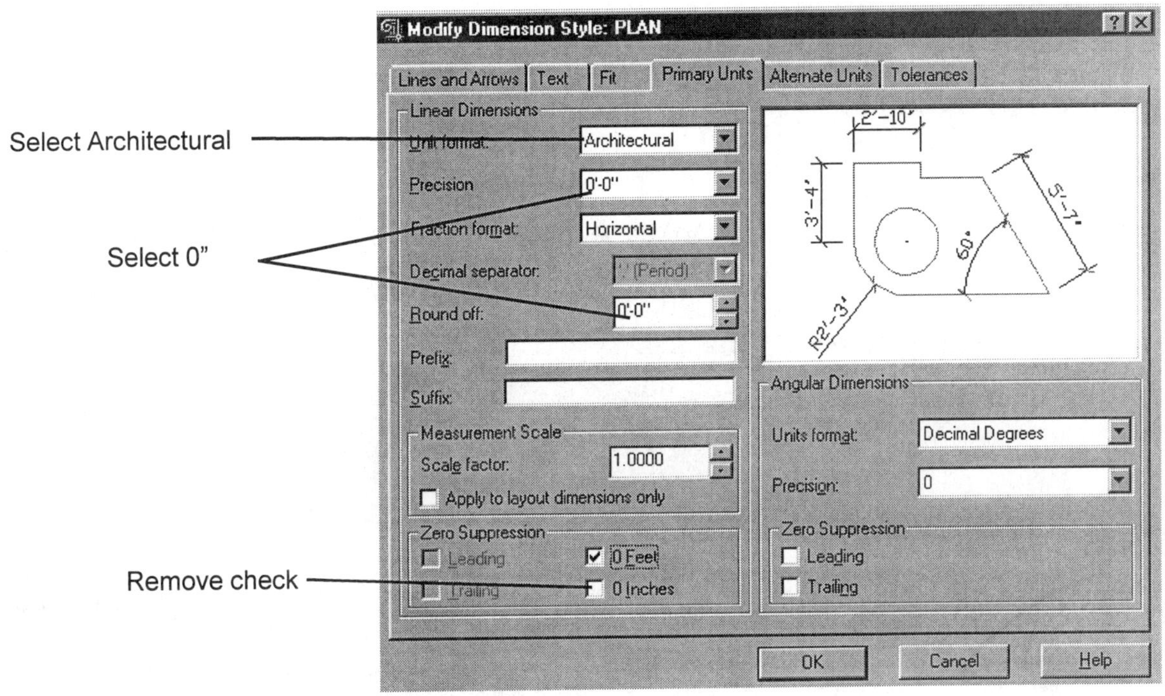

Fig. 10-5 Change settings of Primary Units.

You selected 0'-0" for round off and precision because you are dimensioning a somewhat large plan and you do not want to get odd measurements such as 15/16." You also removed suppression of inches so that whole measures such as 57' will appear as 57'-0."

12. Click "OK" to exit New Dimension Style window.
13. In Dimension Style Manager window, highlight "PLAN" and click "Set Current" to make PLAN the current dimension style.
14. Click "Close" to exit.

This dimension style was set up to plot the floor plan at 1/4"=1' scale. If you want to plot the drawing to 1/8"=1" scale, you would double the value of all the settings (12" for text height, 6" from text offset, etc.) For 1/2"=1' scale, you would divide the values by one-half (3" for text height, etc.).

With your dimension style defined and made current, you can dimension your floor plan.

1. **Make sure ORTHO is on.**
2. **Set SNAP to 1 inch.**
3. **Menu Bar: Tools > Drafting Settings > Object Snap Tab.**
4. **Check in the box next to "intersection" until a check appears.**
5. **Click on "OK" to exit dialogue box.**
6. **Turn OSNAP on.**

The dimensioning commands are easily accessed by using toolbars or Dimension on the menu bar. Try using both methods when doing dimensions.

1. **Menu Bar: View > Toolbars.**
2. **Check "Dimension" and click "Close" Button.**

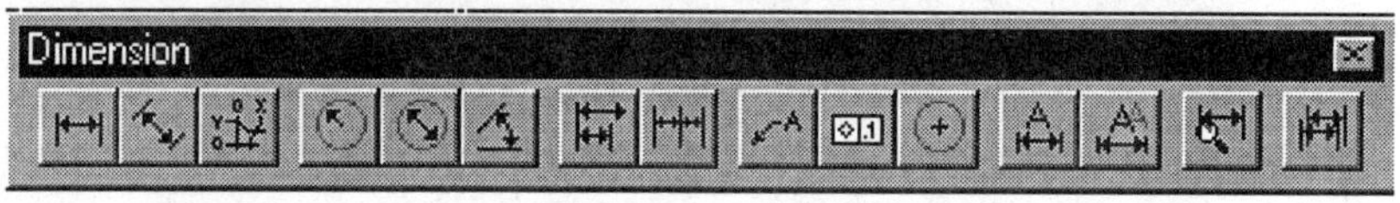

DIMLINEAR Command

Linear dimensioning is the most common type used in building design. The DIMLINEAR command allows you to draw dimensions in any horizontal, vertical, or rotated direction. Refer to illustration (Fig. 10-6).

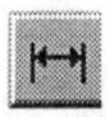

Command: **DIMLINEAR**
Specify first extension line origin or <select object>: **(point A)**
Specify second extension line origin: **(point B)**
Specify dimension line location or
[Mtext/Text/Angle/Horizontal/Vertical/Rotated]: **(point C)**

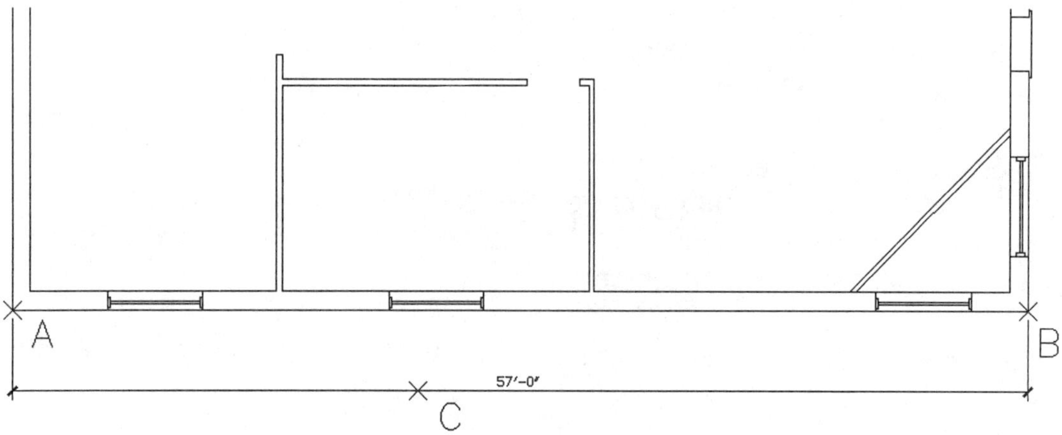

Fig. 10-6 Use of the DIMLINEAR command.

Notice that the point for the dimension line location was below the extension line. Had you wanted it above, you would have selected a point above the location points for the extension lines.

DIMLINEAR Command - Rotate Option

Now you will draft the dimension line for the planter wall that is on an angle (Fig. 10-7). Because you drafted the wall on a 45 degree angle, it will use the ROTATE option and rotate the dimension line 45 degrees. The dimension text on your plan may vary from the illustration.

1. Turn ORTHO off.

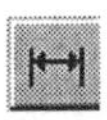

Command: **DIMLINEAR**
Specify first extension line origin or <select object>: **(point A)**
Specify second extension line origin: **(point B)**
Specify dimension line location or
[Mtext/Text/Angle/Horizontal/Vertical/Rotated]:**R**
Specify angle of dimension line <0>: **45**
Specify dimension line location or
[Mtext/Text/Angle/Horizontal/Vertical/Rotated]: **(point C)**

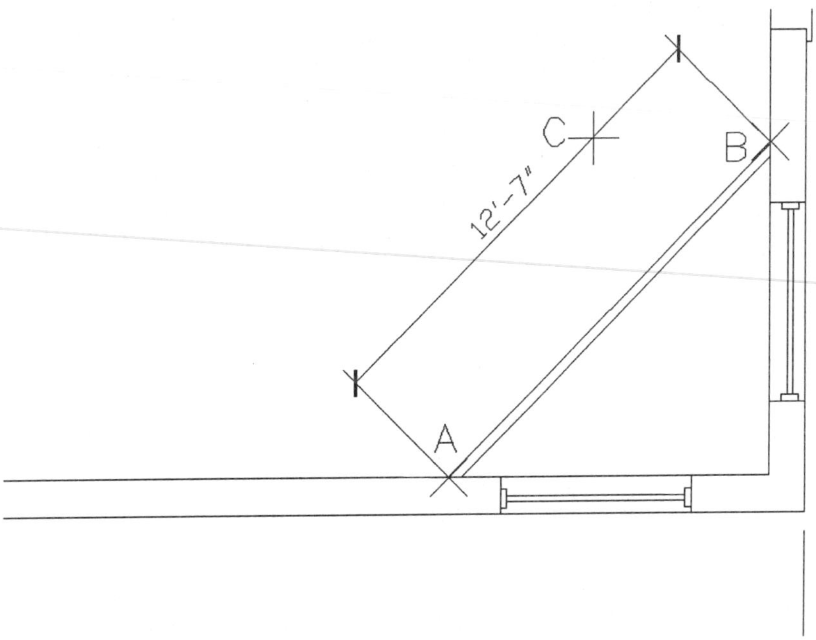

Fig. 10-7 Rotating a dimension.

DIMCONTINUE Command

This command allows you to continue dimensions along the same dimension line as you select for the first one (Fig. 10-8). You may want to use the scroll bars to pan so that you can select the lines easier.

1. Turn ORTHO on.
2. Click on OSNAP on status line to turn on.

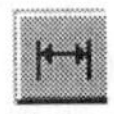

Command: **DIMLINEAR**
Specify first extension line origin or <select object>: **(point A)**
Specify second extension line origin: **(point B)**
Specify dimension line location or
[Mtext/Text/Angle/Horizontal/Vertical/Rotated]: **(point C)**

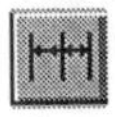

Command: **DIMCONTINUE**
Specify a second extension line origin or [Undo/Select] <Select>: **(point D)**
Specify a second extension line origin or [Undo/Select] <Select>: **(point E)**
Specify a second extension line origin or [Undo/Select] <Select>: **(point F)**
Specify a second extension line origin or [Undo/Select] <Select>: **(point G)**
(press ENTER)
Select continued dimension: **(press ENTER or right click)**

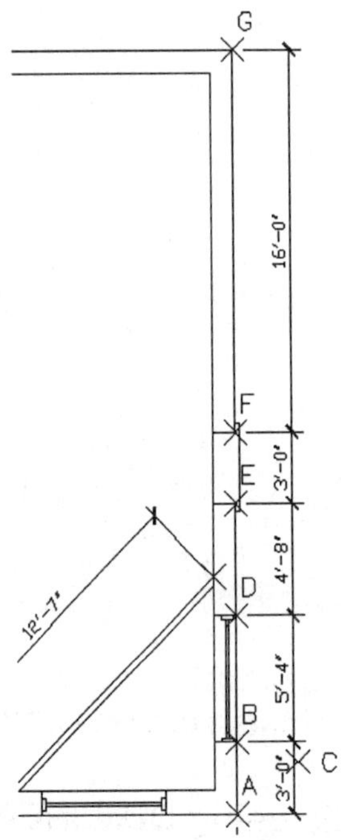

Fig.10-8 Continued dimensioning.

Dimensioning the Interior Walls

You will finish dimension the interior and exterior walls. (Fig.10-9):

1. **Menu Bar: Tools > Drafting Settings...**
2. **Click Object Snap Tab.**
3. **Click "Clear All" button.**
4. **Check "Perpendicular" and "OK" to exit.**
5. **Freeze A-DOOR layer.**

QDIM Command

This command will quickly dimension a line or other object. The 9'-6" and 11'-2" vertical dimensions were done using this command (Fig. 10-9).

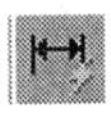

Command: **QDIM**
Select geometry to dimension: **(select a line)**
Select geometry to dimension: **(press ENTER)**
Specify dimension line position, or [Continuous/Staggered/Baseline/Ordinate/ Radius/Diameter/datumPoint/Edit] <Continuous>: **(point)**

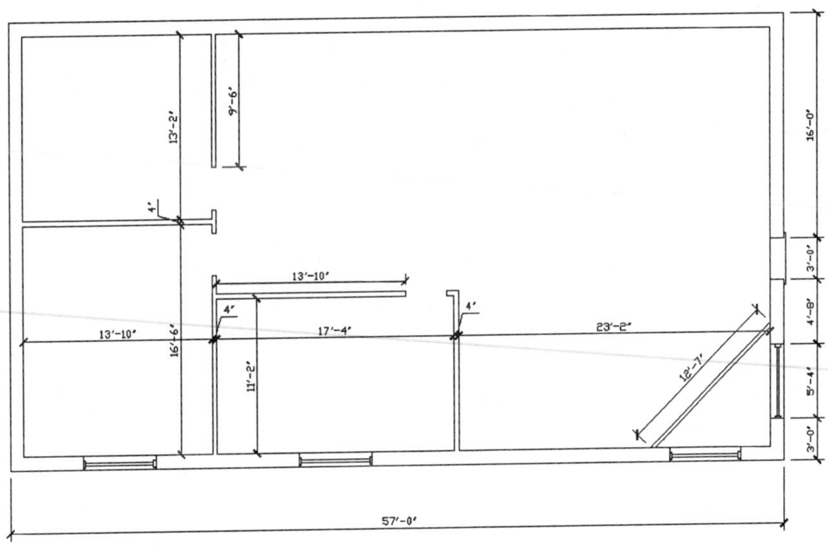

Fig. 10-9 Finish dimensioning.

1. **Finish dimensioning walls using DIMLINEAR command and similar to illustration (Fig. 10-9).**
2. **Thaw the A-DOOR layer.**

Modifying Dimensions

If your style is not correct, simply modify the Dimension Style and make it current. If dimensions do not change, use Dimension Update. You can also use Update Dimension to set dimensions to a new style as long as that style is current.

The DDMODIFY command allows you to change height of the dimension text or individually change arrowhead size as well as other values indepedent of the dimension style.

If you want to change the text only, use the DDEDIT command to go directly to the text editor. Or, while you are dimensioning with DIMLINEAR command, type T to type different text than what AutoCAD lists as default value.

The easiest way to change the dimension line location is to use the STRETCH command making a crossing window over both the extension and dimension lines.

As a last resort, you can EXPLODE the dimension line but this will discard the dimension style so you cannot update it. The exploded dimension will be as if you had done it with the LINE and MTEXT commands.

DIMTEDIT Command

You can move the text by using the DIMTEDIT command.

1. Toggle ORTHO off.

Command: **DIMTEDIT**
Select dimension: **(select a dimension)**
Enter text location <Left/Right/Home/Angle>: **(move to new location)**

AutoCAD Release 14 Files and Dimension Styles

Dimension style and the Dimension Style window were changed extensively in AutoCAD 2000. A Release 14 drawing file will list different values for dimension styles with the same appearance on the screen. However, once you establish a new dimension style, the values in this tutorial will work.

LEADER Command

This command can be used to note any drawing. In the following sequence, you will use a "spline" to make the note clearly different from the dimension line. As with other dimensioning it can be accessed through the pulldown menu "Draw" (Fig. 10-10).

1. Turn ORTHO off by pressing the F8 key or clicking on ORTHO on status line.
2. Menu Bar: Dimension > Leader.

_qleader
Specify first leader point, or [Settings]<Settings>: **S**

1. Click on Leader Lines & Arrows Tab.
2. Check "Spline" box "OK" to close window.

Specify first leader point, or [Settings]<Settings>: **(point A)**
Specify next point: **(point B)**
Specify next point: **(point C)**
Specify next point: **(press ENTER)**
Specify text width <0">: **(press ENTER)**
Enter first line of annotation text <Mtext>: **BRICK PLANTER**
Enter next line of annotation text: **(press ENTER)**

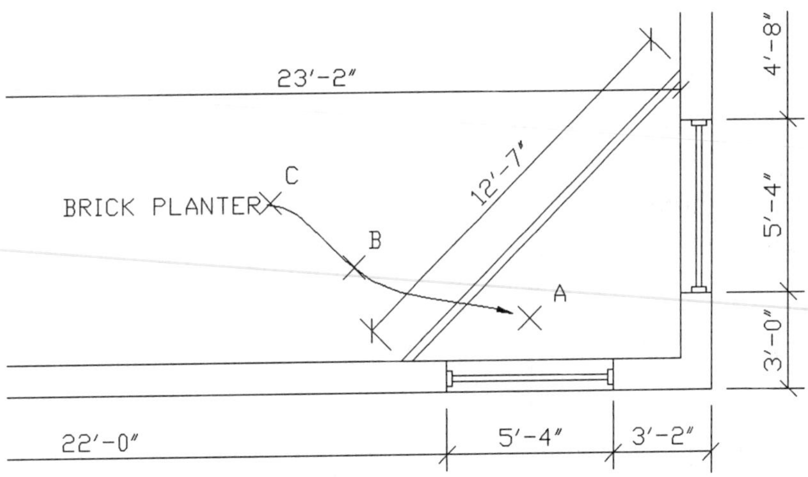

Fig. 10-10 Placing a leader note on your plan.

1. Use the MOVE command selecting only the text to move leader.

The spline adjusts to the new location while keeping the endpoint in its same location.

Drafting and Dimensioning Cabinet Sections

This section will take some time--you may want to save it for reference later.

1. Menu Bar: File > New.
2. Click on "Start from Scratch" and then "OK" to exit.
3. Set up the drawing:

UNITS	**Architectural**
LIMITS	**upper right corner 22',17'**
GRID	**3"**
ZOOM	**all**

Look at the illustration first before you set up the dimension to see how you will draft and dimension the bar section (Fig. 10-11).

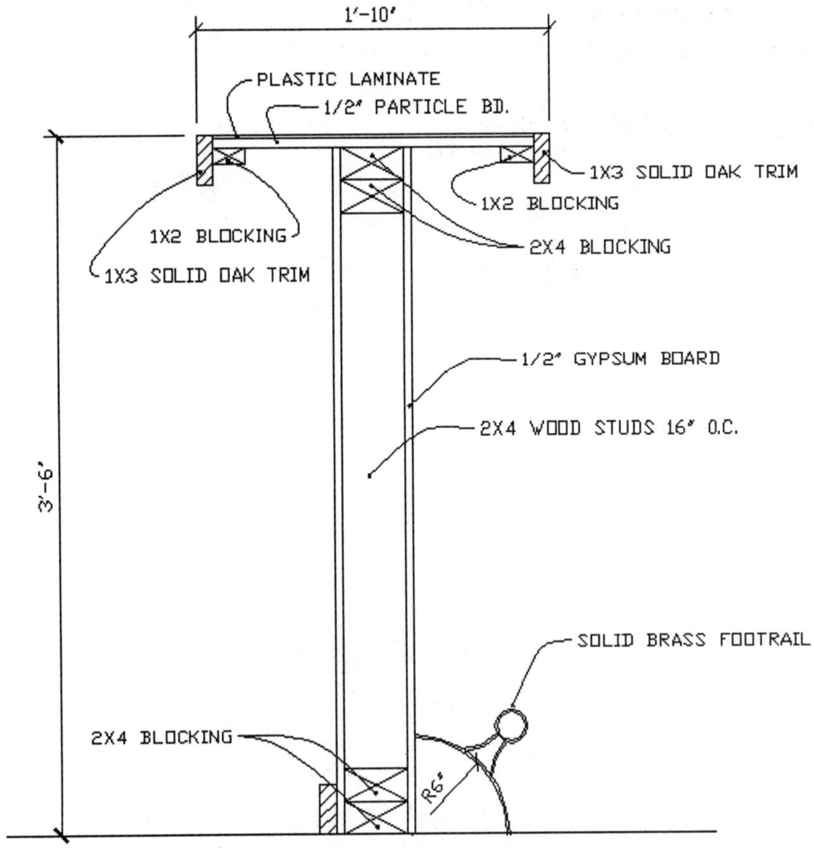

Fig.10-11 Final section of bar.

First, you should draft the actual bar section without dimensions or notes although you will need to refer to the illustration for reference on sizes (Fig. 10-11). You can draft it many different ways but below is listed one method.

1. Use the SNAP command and set to 1/4".
2. First draft the floor line 20" wide with the LINE command.
3. Use OFFSET command to offset the floor line 3'6" in order to get the top line which you can offset to get the plastic laminate and the 1/2" particle board.

TIP: Next you can make the two 2x4 blocks. If you are doing a number of different sizes of blocking, 2 by 4, 1 by 2, or 2 x 6 for instance, it is easiest to make a 1 x 1 block and use the X and Y scale to size the blocking. For a 2 x 4 block laid horizontally for example, you would use 4 for the X scale and 2 for the Y scale.

1. Insert the two 2x4's inserting to the midpoint of the floor line.
2. Draft the two lines for the studs using the EXTEND command to extend the lines to the particle board.
3. Draft the oak trim using a diagonal hatch.
4. Draft the footrail using ARC command with a 6" radius.

Your cabinet should look similar as illustrated (Fig. 10-12).

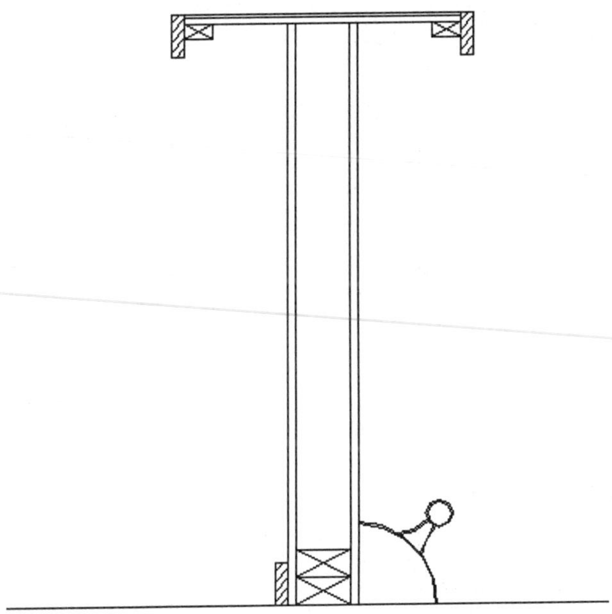

Fig. 10-12 Bar section without dimensioning or notes.

The next step is to set up the dimension style. The following values will make a dimension style that will plot with text at 1/8" height when plotted 2"=1'-0" scale.

1. Menu Bar: Format > Dimension Style...
2. Set up the following setting the values as indicated for each tab:

Dimension Style Name: CABINET

Lines & Arrows Tab:
Arrowheads:
1st: Architectural tick
Size: 3/4"
Extension Lines:
Extend beyond dimension lines: 3/4"
Offset from origin: 3/4"

Text Tab:
Height: 3/4"
Offset from dimension line: 3"
Vertical: Above
Text Alignment: Aligned with dimension line

Primary Units Tab:
Architectural
Round Off and Precision: 0'-0"
Remove check from Inches in Zero Suppression area

1. Make the two dimensions for the top and height.

DIMRADIUS Command

The DIMRADIUS command will indicate the radius of an object. The radius of the footrail was done with this command:

Command: **DIMRAD**
Select arc or circle: **(select arc)**
Dimension text = 6"
Specify dimension line location or [Mtext/Text/Angle]: **(point along arc to where you want text located)**

The Text option allows you to enter different text such as "Rad=6" or other text.

Next, you will use the LEADER command to enter the notes on materials. The following shows the procedure for one of the materials.

1. Menu Bar: Dimension > Leader.

_qleader
Specify first leader point, or [Settings]<Settings>: **S**

1. Click on Leader Lines & Arrows Tab.
2. For arrow head, select "dot small."
3. Check "Spline" box "OK" to close window.

Specify first leader point, or [Settings]<Settings>: **(point)**
Specify next point: **(point)**
Specify next point: **(point)**
Specify next point: **(press ENTER)**
Specify text width <0">: **(press ENTER)**
Enter first line of annotation text <Mtext>: **2x2 BLOCKING**
Enter next line of annotation text: **(press ENTER)**

On the notes where there are two arrows, repeat the LEADER command and simply press ENTER key when the text prompt comes up and then click "OK" to exit editor to enter no text. If you do not like where you placed a note, use the MOVE command. If you select the text only, the dot stays where it was but the spline will move to the new location.

This procedure will make a style override to the cabinet dimension style. You will save it to the cabinet style:

1. Menu Bar: Format > Dimension Style.
2. Right click on <style override> and select "Save to current style."
3. Click on "Close" to exit.

This will only change the leader arrowhead style to dot small. The dimension line arrows will still be architectural ticks.

Practice Dimensioning

This tutorial, while giving you a good introduction to dimensioning, is by no means complete. You will want to practice with dimensioning to become fully proficent with automatic dimensioning. The references at the end of this book will also provide more detail on dimensioning.

Template or Dwt Files

Establishing dimensioning styles is certainly not a task you would want to do on every drawing. The smart procedure is to have drawings set up with your most commonly used dimension styles, text styles, and typical title blocks. You can save these files as templates by using "Save as" and then clicking on "Drawing Template File (*.dwt)" for the file format. When starting a new drawing, just click on "Use a Template" on the "Create New Drawing" dialogue box and use your template file.

TUTORIAL 11

DRAWING LAYOUT WITH PAPER SPACE

Commands Learned:	TILEMODE	VPORTS	PSPACE
	MVS	MSPACE	

So far, you have been drawing or drafting in model space. AutoCAD also has paper space which allows you to open viewports into your drawing. Think of it as a piece of tracing paper laid over the drawings. When you make your viewports, you are cutting holes into the tracing paper to view the drawings in model space. By zooming model space in relation to paper space, you bring the view of the drawings in either closer or farther away depending on the desire scale of the final print.

1. **Open the drawing file from the last tutorial.**
2. **Thaw the A-FURN, A-DOOR, PLANTER, and A-FLOOR layers.**
3. **Make the A-FURN the current drawing layer.**
4. **Freeze the A-WALL-DIMS layer.**
5. **Use the ZOOM command**  **to zoom all of the drawing.**

The display should appear as in the illustration (Fig. 11-1).

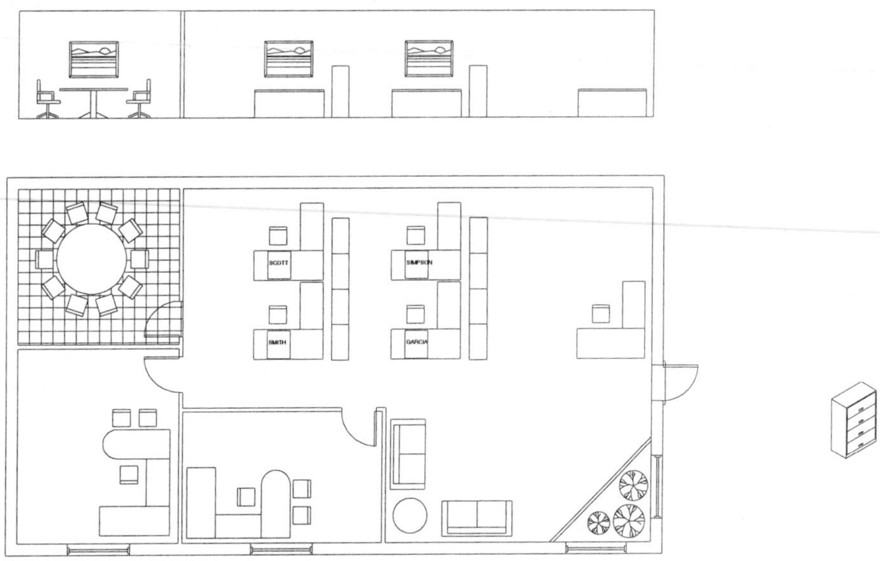

Fig. 11-1 View of drawing.

TILEMODE Command

You enter paper space by turning tiled viewports off. This is done by setting the system variable TILEMODE to 0 or by clicking on the Layout1 button.

Command: **TILEMODE**
New value for TILEMODE <1>: **0**

PAGESETUP Command

The "Page Setup - Layout1" window appears. This is similar to a print or plot window except you do not have a preview nor does it send it to the printer when finished. PAGESETUP is also on the Menu Bar under File. Layout1 is the Layout name which will be your first page. To the right, the "Add..." button allows you to make new layouts.

1. Click Plot Device Tab.
2. Select a D size plotter (e.g. HP DesignJet) or None if you do not have a large size plotter.
3. Click Layout Settings.
4. Select ANSI E (36.00 x 44.00 Inches) for paper size.
5. Check that inches, Landscape for orientation, and Layout for Plot area are checked and the scale is 1:1.
6. Click "OK" to exit "Page Setup" window.

The screen shows a white page indicating the paper size or layout (Fig. 11-2). It has a dashed outline which indicates the plotting area; this is not a part of the drawing. The white page is against a gray background. One viewport has already been placed within the white layout page. However, you will erase it because you want to make smaller viewports.

1. Erase the existing viewport by clicking on the outline (Fig. 11-2).

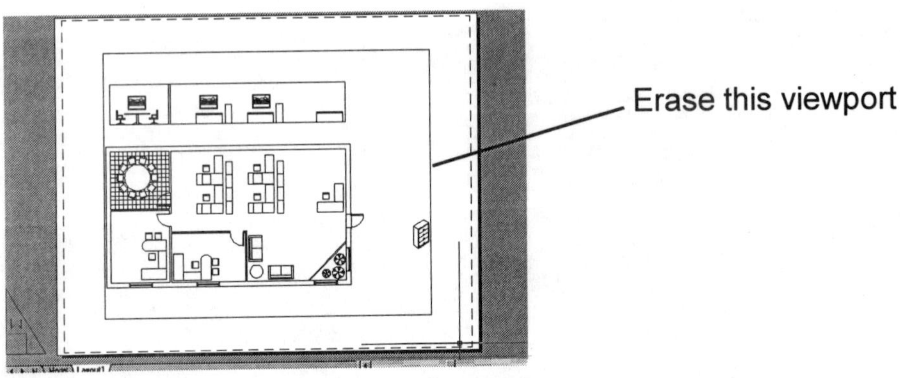

Fig. 11-2 Erase the existing viewport.

1. Make two new layers with assigned color values:

A-VPORT	**blue**
A-TTLB	**magenta**

2. Make A-TTLB the current layer.

3. Turn SNAP and ORTHO off.

MVSETUP Command

The MVSETUP command allows you to insert a titleblock or make viewports.

Command: **MVSETUP**
Enter an option [Align/Create/Scale viewports/Options/Title block/Undo]: **T**
Enter title block option [Delete objects/Origin/Undo/Insert] <Insert>: **(press ENTER)**
Available title blocks:...

0:	None
1:	ISO A4 Size(mm)
2:	ISO A3 Size(mm)
3:	ISO A2 Size(mm)
4:	ISO A1 Size(mm)
5:	ISO A0 Size(mm)
6:	ANSI-V Size(in)
7:	ANSI-A Size(in)
8:	ANSI-B Size(in)
9:	ANSI-C Size(in)
10:	ANSI-D Size(in)
11:	ANSI-E Size(in)
12:	Arch/Engineering (24 x 36in)
13:	Generic D size sheet (24 x 36in)

Enter number of title block to load or [Add/Delete/Redisplay]: **13**
Enter an option [Align/Create/Scale viewports/Options/Title block/Undo]: **(press ENTER)**

A generic D-size title block is added to the drawing (Fig. 11-3).

Fig. 11-3 View after title block insertion.

1. **Change SNAP to 1/4.**
2. **Change the current drawing layer to A-VPORT.**
3. **Zoom to drawing extents.**

You can think of paper space as a large piece of paper in front of the model space. In order to see into model space, you must first cut out windows or viewports. So, to view your drawings, you will be opening up viewports and at the same time will be returning to model space. There will be one viewport on top to place your elevation, and two on the bottom to place the floor plan and isometric drawing. Refer to points in the illustration (Fig. 11-4) to make the boundary for a single viewport for the elevation:

Command: **MVS**
Enter an option [Align/Create/Scale viewports/Options/Title block/Undo]: **C**
Enter option [Delete objects/Create viewports/Undo] <Create>: **(press ENTER)**
Available layout options: . . .
0: None
1: Single
2: Std. Engineering
3: Array of Viewports
Enter layout number to load or [Redisplay]: **1**
Specify first corner of bounding area for viewport(s): **(window near top but inside of title block)**
Enter an option [Align/Create/Scale viewports/Options/Title block/Undo]: **(press ENTER)**

Fig. 11-4 Insertion points for viewport array.

VPORTS Command

You can also insert viewports with the VPORTS command. In addition, you can make nonrectangular viewports or remove hidden lines from three-dimensional views in viewports.

Command: **VPORTS**
Specify first corner or [Fit] <Fit>: **(make a viewport similar to Fig. 11-5).**

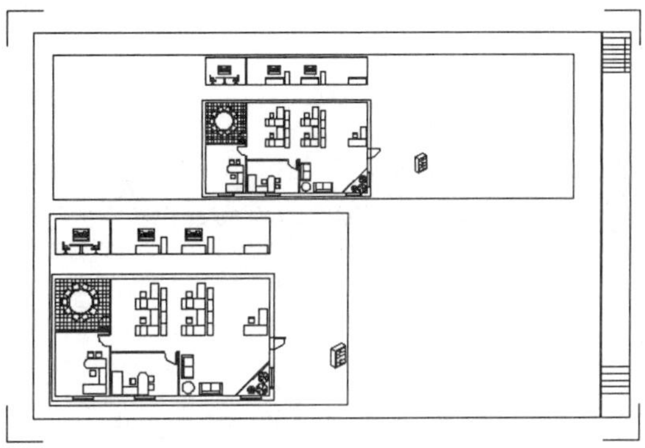

Fig. 11-5 Two viewports to be used for elevation and furniture plan.

You will use VPORTS with the "-" placed before the command so that you will see all the options at the command prompt:

Command: **-VPORTS**
Specify corner of viewport or
[ON/OFF/Fit/Hideplot/Lock/Object/Polygonal/Restore/2/3/4] <Fit>: **P**
Specify start point: **(point for first point of polygon)**
Specify next point or [Arc/Close/Length/Undo]: **(continue pointing and use C to close polygon as illustrated Fig. 11-6).**

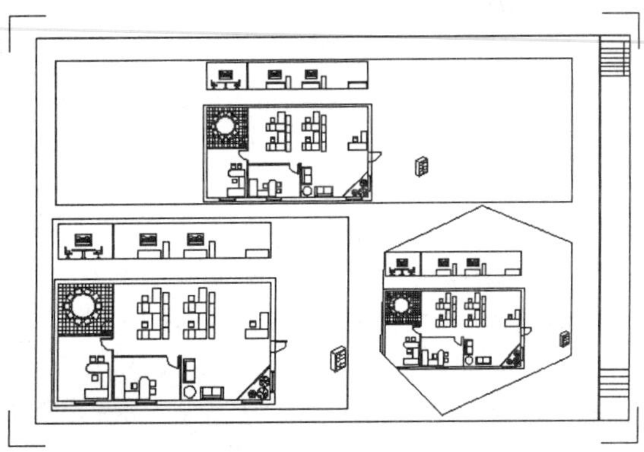

Fig. 11-6 Making a non-rectangular viewport.

The “Polygonal” option works well for an L-shaped floor plan too. The Object option may also be useful. You first must make the object with a draw command (e.g. CIRCLE, POLGON, ELLIPSE, etc.). Then use -VPORTS and type “O” to select the object to use for the shape of the viewport.

MSPACE Command

By clicking on the Model tab next to Labout1, you return to model space or set tilemode back to 1.

1. **Click on the Model space tab.**
2. **Turn on the UCSICON if you do not see the UCS icon in the lower left corner.**
3. **Turn off the GRID by clicking on the F7 key.**

You see your drawing as before enabling you to easily edit your drawings or fix errors. Assuming that your drawing is correct, you will finish arranging your layout by scaling the viewports. This task is accomplished by being in paper space layout and using the MSPACE command to enable model space.

1. **Click on the Layout1 tab.**

Command: **MSPACE**

The crosshairs do not appear over the whole paper, only within a viewport. By clicking in each viewport, you can make it current to scale the viewport.

1. **Click on the lower left viewport to make it current.**
2. **PAN**  **so that the floor plan is in the center of the viewport.**

You will be using the XP option of the ZOOM command to scale your drawings in relation to paper space. With the lower left viewport as the current drawing viewport, enter the following to the command prompt:

Command: **Z**
ZOOM
Specify corner of window, enter a scale factor (nX or nXP), or
[All/Center/Dynamic/Extents/Previous/Scale/Window] <real time>: **1/48XP**

Your floor plan will be 1/4"=1' scale when plotted out on 24" x 36" paper. The plan may not quite fit in your viewport—do not worry because we can fix that later by stretching the viewport. Next, you will make a 1"=1' view of the lateral file.

1. **Click on the lower right viewport.**

2. ZOOM with a window  **to view the lateral file only.**

Command: **Z**
ZOOM
Specify corner of window, enter a scale factor (nX or nXP), or
[All/Center/Dynamic/Extents/Previous/Scale/Window] <real time>: **1/12XP**

1. PAN to make lateral file fit in right side of viewport.

Next, you will make the elevation 1/2"=1' scale.

1. Click on top viewport and make sure the elevation is in full view, without the floor plan, by using the ZOOM command with a window.
2. Turn off the GRID (F7 key).

Command: **Z**
ZOOM
Specify corner of window, enter a scale factor (nX or nXP), or
[All/Center/Dynamic/Extents/Previous/Scale/Window] <real time>: **1/24XP**

Your views should look similar to the illustration (Fig. 11-7).

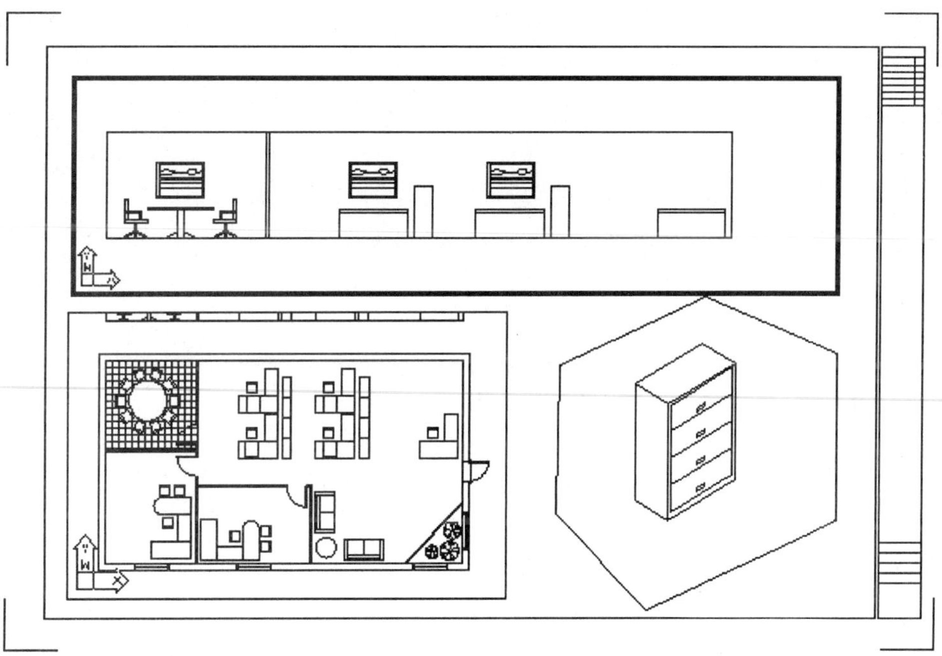

Fig. 11-7 Viewports after scaling to paper space

In the floor plan viewport, you see part of the elevation. You will freeze the A-FURN-ELEV only in the floor plan viewport. If you just froze the layer, it would affect all viewports.

1. Make sure you are in model space and the floor plan viewport is current.

Command: **LA**

2. Highlight the A-FURN-ELEV layer.
3. Click to make check in box next to "Freeze in active viewport" (Fig. 11-8).

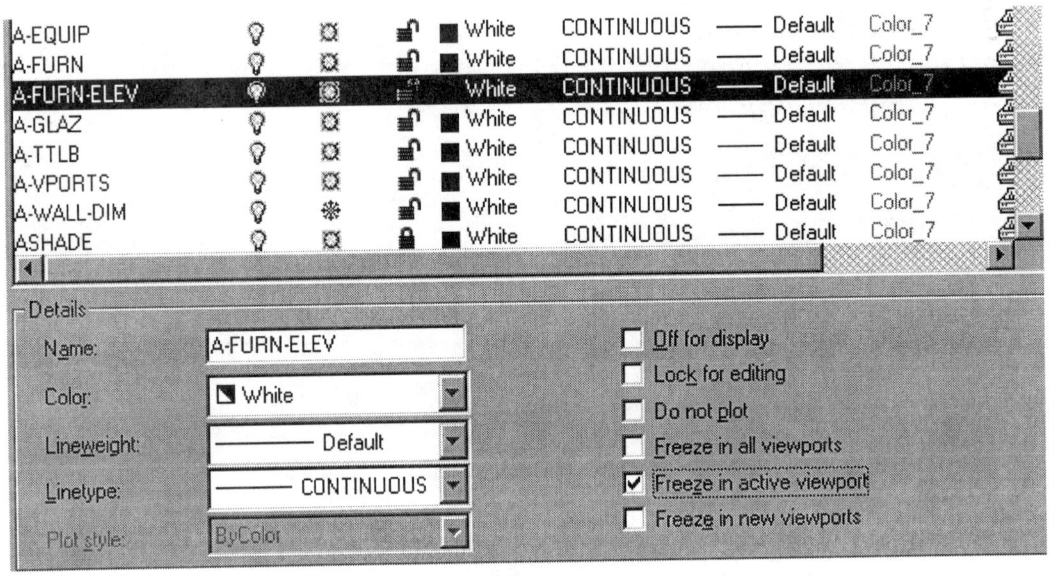

Fig. 11-8 Freeze layer in active viewport.

The elevation should disappear. You also could pan in the viewport until the elevation disappears. Panning does not change the magnification or zoom.

If you can visualize adding the lettering in paper space, you can see that the layout of the drawings is not good. There are several ways you can line up the drawings. The easiest is to draw construction lines in paper space using the LINE command to center and move the viewports to line up (Fig. 11-9). In the illustration "apparent intersection" was used to find the lower right corner of the elevation to use for first point of the construction line. In addition to making construction lines, you will change the crosshair size to 100% so that you can use the crosshairs for lining up. You can zoom into a small area but make sure you are in paper space. If you were in model space and zoomed in, you would be required to redo the zoom in relation to paper space.

1. Click on the "MODEL" button on the status line until it says PAPER or type PSPACE to command prompt to return to paper space.
2. Menu Bar: Tools > Options...
3. Click on Display tab and change crosshair size to 100.
4. Use the LINE command to draw a 4" construction line, make a copy, and place on both sides of elevation.
5. Center the elevation.
6. Draw a line to line up lateral file with elevation and another line to line up elevation with floor plan.
7. Move viewports by clicking on them to line up drawings similar to illustration (Fig. 11-9).

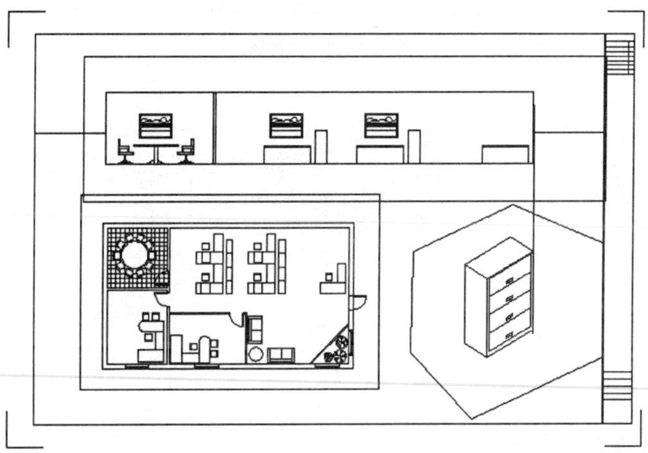

Fig. 11-9 Aligning drawings with construction lines.

1. Erase the construction lines you made to line up the drawings.

You can also adjust the size of the viewports by stretching them using a crossing window to select both the top and vertical line of the viewport. Keeping ORTHO on will insure the viewport stretch either horizontally or vertically.

You will add the lettering and freeze the A-VPORTS layer so the outline of the viewports does not print or plot. Text will be entered in actual heights. That is, 1/4" height will be 1/4" when your drawing is plotted out at 1=1 or full scale.

TIP: When you make one drawing label, make a block of the label to copy to the other two drawings. Then use DDEDIT to change text.

1. Make the A-TTLB layer the current drawing layer.
2. Freeze A-VPORTS layer (do not use "Freeze in active viewport").
3. Menu Bar: Format > Text Style...
4. Make a new text style for the titleblock text.
5. Make a 3/4" diameter circle and a line for the drawing labels.
6. Use DTEXT command for drawing names at 1/4" height and scale at 1/8" height and drawing letter is 1/2" height.
7. Use DTEXT command to enter "Sales Office" rotating 90 degrees and entering 3/4" height while rest of text is 1/4" height.

Once again, it would be smart to save this file as a template for D size drawings. You would already have the dimension style, text styles, some office planning blocks, and viewport layout. Alternatively, you could use the Design Center, open this drawing file, and drag these features to a new drawing (See Tutorial 7).

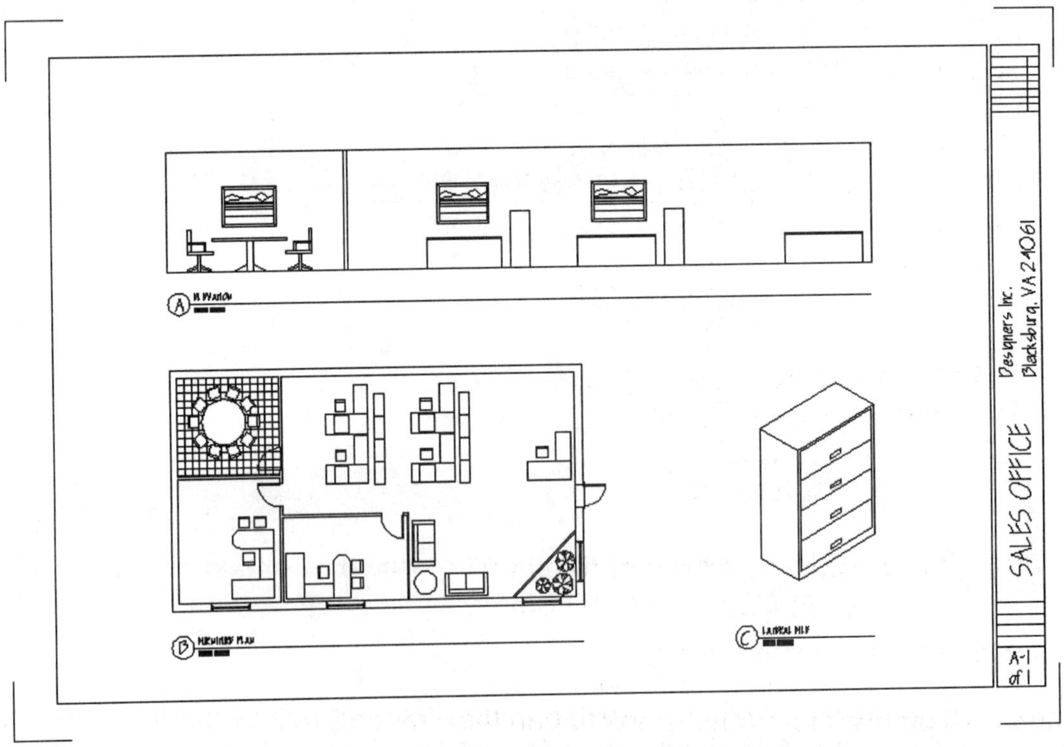

Fig. 11-10. Layout of first page.

Linetype Scale and Paper Space

When you did the lighting plan, the switching was done using a dashed line assigned to the layer. You then changed the LTSCALE to 10. If you print this out, the dashed line would appear solid. Because the lighting plan will be printed to 1/4"=1'-0" scale and paper scale is 1=1, the LTSCALE should be changed to .25 (1/4 x 1). Paper scale does not affect the scale of hatch pattern though.

This drawing could be printed out by using a scale of 1=1 or on a letter-size printer by using "Scaled to Fit." You may want to print now since the next section will change the display of layers on this page.

Using the Layout Wizard

AutoCAD 2000 provides a Layout Wiizard to create paper space layouts quickly. However, it is still important for you to be proficient with MVSETUP and VPORTS commands in case you need to change the layout adding additional viewports. You will use the Layout Wizard to create a second page containing the floor plan with dimensions and the lighting plan. Note the illustration to see how the page will appear (Fig. 11-11).

1. Thaw the A-VPORTS layer and make A-TTLB layer current.
2. Menu Bar: Tools > Wizards > Create Layout...
3. Enter the following specifications and click "Next >" button:

Name for new layout: Layout2
Printer: None if no large plotter (e.g. HP DesignJet 750C)
Paper Size: ANSI E (34.00 x 44.00 Inches)
Orientation: Landscape
Title Block: Generic 24in x 36in

Define Viewports:

Viewport Setup: Array	**Viewport Scale: 1/4"=1'0"**
Rows: 2	**Space between rows: 1**
Columns: 1	**Space between columns: 1**

Pick Location: Click "Select location" button and make a window which almost fills inside of title block.
4. Click Finish button.

Now you will go back to model space to pan the views as well as put the dimension plan in the bottom viewport and the lighting plan in the top viewport.

1. Zoom extents.
2. Click on Paper icon on status line until it says Model (or enter MSPACE at command prompt).
3. Click in top viewport to make it active.
4. Pan to see floor plan.
5. Use LAYER command and freeze the following layers in the active viewport so you will only see lighting plan (note that you will have to thaw layers first):
    - A-FURN
    - A-FURN-ELEV
    - A-FURN-ISO
    - A-WALL-DIMS
    - PLANTER
6. Click in bottom viewport and pan to see floor plan.
7. Use LAYER command and freeze the following layers in the active viewport so you will only see floor plan with dimensions:
    - A-CEILING-GRID
    - A-DOOR
    - A-FURN
    - A-FURN-ELEV
    - A-FURN-ISO
    - E-LITE
    - E-LITE-SWITCH
8. Click on Model on status line until it says Paper (or type PSPACE at command prompt).
9. Arrange viewports.
10. Use DDMODIFY to change viewports to A-VPORTS layer.
11. Freeze A-VPORTS layer.
12. Click on Layout1.
13. Make a WBLOCK of the drawing label for the furniture plan.
14. Make a WBLOCK of text only in title block using the lower right corner of titleblock as insertion point.
15. Click on Layout2 and insert blocks you just made.
16. Explode the blocks and use DDEDIT to change page number and drawing label names.
17. Change LTSCALE to .25 value.

Your final second page may look similar to the illustration (Fig. 11-11). The page would make a better layout if it had a lighting legend or key and perhaps some notes to balance the page; however, that is not necessary for this tutorial.

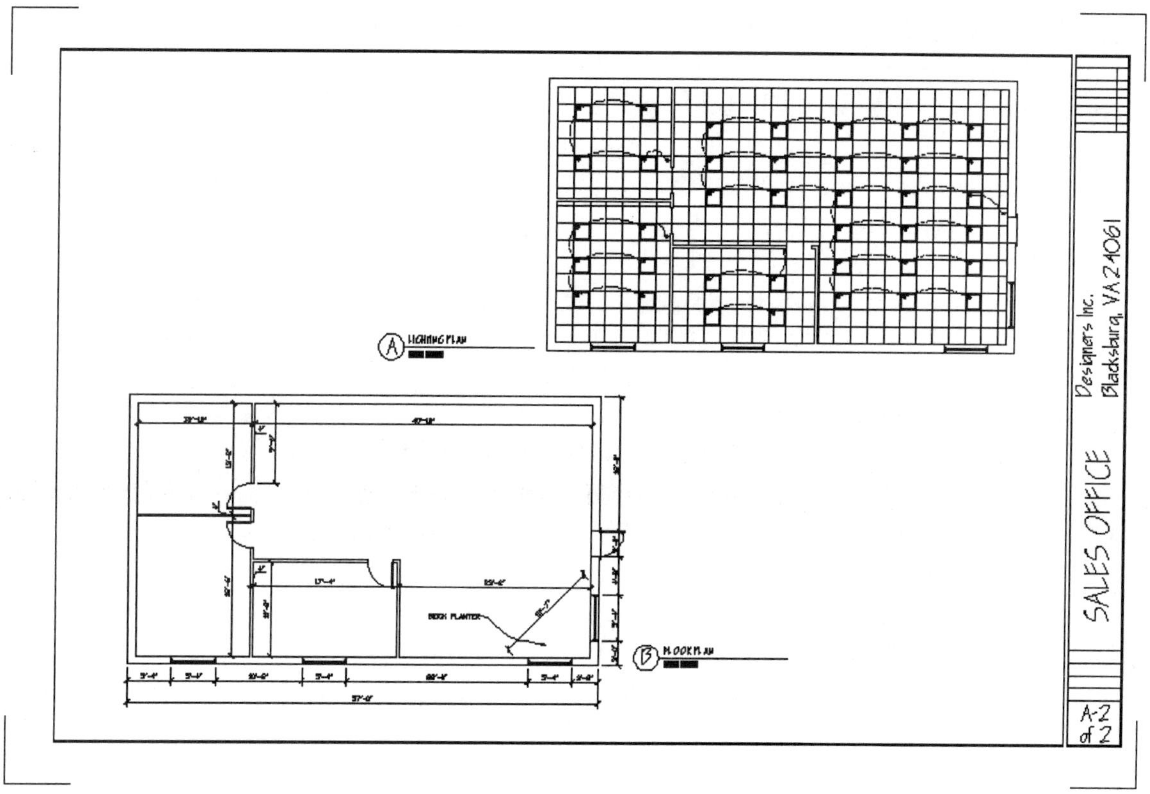

Fig. 11-11 Using the Layout Wizard.

If you wanted to go back and print Layout1, you would need to freeze layers in active viewports again because dimensionsing and lighting layers are displayed.

This is the end of this tutorial. You can print Layout1 and Layout2 on a letter-size printer by using "Scaled to Fit" for scale. You can also plot this drawing on a D or larger size plotter by using the plot setting parameters at the end of the Drawing Procedure section on the next two pages.

The following section is a general drawing checklist and procedure that could be used as a reference for CAD projects.

Drawing Checklist

- **Drawing Setup: Units, Limits, Grid, Zoom all, Layers:**

Layer Name	Drawing Elements	Drawing Area
A-WALL	Walls	Model Space
A-GLAZ	Windows	Model Space
A-DOOR	Doors	Model Space
A-WALL-DIM	Plan Dimensions	Model Space
A-ELEV	Elevations, Elev. Dimensions	Model Space
A-FURN	Furniture, other Labeling inside plan	Model Space
A-VPORT	Viewports	Paper Space
A-TTLB	Titleblock, Drawing Labels, Titleblock text, Elev. Symbols	Paper Space

- **Dimension Style:**

PLAN

Lines & Arrows Tab:

Arrowheads:

Architectural tick:

Size: 6"

Extension Lines:

Extend beyond dimension lines: 6"

Offset from origin: 3"

Text Tab:

Height: 6"

Offset from dimension line: 3

Vertical Placement: Above

Text Alignment: Aligned with dimension line

Primary Units Tab:

Architectural

Round Off and Precision: 0'-0"

Remove check from Inches in Zero Suppression area

Note: The above values are for dimensioning to plot out at 1/4"=1'-0" scale. Use the following formulas for other plot scales:

1/2"=1'-0"	divide the above values by 2 (e.g. 3", 1-1/2")
1/8"=1'-0	multiply the above values by 2, (e.g. 12", 6")
1"=1'-0"	divide the above values by 4 (e.g. 1-1/2", 3/4")

The next section details a suggested drawing procedure.

Drawing Procedure

1. Use PLINE to draft outer wall.
   Use OFFSET to offset polyline 9"
   EXPLODE polylines.
   Draw interior walls with either PLINE and offset 6" or just use LINE command. Use TRIM and extend to make sure lines connect. Break out walls with TRIM and/or BREAK command. Draw furniture and other detail.

   Labeling inside plan in model space:

Text	Text height	Plot Scale	Actual Plot Size
Room Names	12"	1/4"=1'-0"	1/4"
Labeling notes	6"	1/4"=1'-0"	1/8

   Note: Use 1/2 the above text height for plans to be printed at 1/2"=1'-0" scale and multiply the text height by 2 for 1/8"=1'-0" scale.

2. Draft elevations by lining up with plan. Any text for notes should be done as above depending on plot scale.

3. Dimension plan and elevations.

4. When finished with drawing scaled elements, use the Layout Wizard to layout your drawings similar to procedure for page 2 in this tutorial. This may take a little experimentation depending on the square footage size of project as well as your desired scale when plotted. You can make multiple layouts if you are not sure which layout works best.

Plot Settings (for D-size plotter):

Paper size:
Plot Area: Display or Extents
Drawing orientation: Landscape
Plot scale: 1=1
Check Center Plot (if using Display)
Full Preview

If what you see on full preview is correct, click on OK to send plot to plotter.

Part Four:
3-D Modeling

TUTORIAL 12

3D ROOM

Commands Learned:	ELEV 3DORBIT PLAN DVIEW HIDE

Up to now, you have been drawing in two dimensions. Now, you will learn how to draw three-dimensional models. AutoCAD provides several ways to draw in three dimensions. These are:

UCS: The User Coordinate System (UCS) allows you to set the drawing plane to any planar surface in a drawing. By default, the UCS is set at WCS or World Coordinate System. The WCS can be thought of as being flat on the ground or floor with 0,0 as the origin in the lower left corner. X is in the horizontal direction while Y is the vertical direction on your screen if you are in plan or top view. In working in 3D, you can change the UCS to any angle or align to an object for drawing.

Elevation: You can set the elevation or how far objects will be elevated from the ground before you make the object with ELEV command. You can think of it as raising the drawing plane. With the DDMODIFY command, you can change the elevation of existing objects.

Thickness: You can set the thickness or height of objects by also using the ELEV command. With the DDMODIFY command, you can change the thickness of some existing objects. You cannot change the thickness or elevation of blocks.

AutoCAD provides two methods—surface modeling and solid modeling—to draw in three dimensions. Whenever possible, you will want to use the surface modeling method because the objects you create will have fewer bytes or require less computer memory than the same objects in solid modeling. Both types of modeling can be used for rendering because you can apply materials to the surfaces created with both types of modeling.

Set up your drawing for 1/4" scale and C size or 18" x 24" paper. The setup should be:

UNITS	**Architectural**
LIMITS	**upper right corner 96',84'**
GRID	**1'**
ZOOM	**All**

1. Save your drawing file.
2. Also make the following layers and assign colors as indicated below:

Layer name	**Color**
columns	**magenta**
3dart	**red**
text	**cyan**
furniture	**green**

1. Use the ZOOM command to see about a 30' x 30' area.

You will be drawing a 20' x 26' x 9' height room with the end wall missing. You will be using an AutoCAD feature, "3D objects," which allows you to draw simple objects easily. You will first use the box command to draw the floor:

1. Menu Bar: Draw > Surfaces.. > 3D Surfaces...
2. In the dialogue box, select the box by clicking on the picture icon and click on OK. Then respond to prompts:

_ai_box
Initializing... 3D Objects loaded.
Specify corner point of box: **5',5'**
Specify length of box: **20'**
Specify width of box or [Cube]: **26'**
Specify height of box: **1'**
Specify rotation angle of box about the Z axis or [Reference]: **0**

1. Use the ZOOM command to see the floor and a little area around.

Now you will draw the west wall but this wall will be set on top of the floor so you will need to set the elevation to one foot before you draw the wall:

ELEV Command

The ELEV command will set the thickness of entities and their elevation. Once you set the elevation or thickness all entities drawn after that point will be drawn with those settings.

Command: **ELEV**
Specify new default elevation <0'-0">: **1'**
Specify new default thickness <0'-0">: **(press ENTER)**

You did not set the thickness because the 3D objects features does that. Now draw the west wall:

1. Menu Bar: Draw > Surfaces.. > 3D Surfaces...
2. In the dialogue box, select the box by clicking on the picture icon and click on OK. Then respond to prompts:

Command: _ai_box
Specify corner point of box: **5',5'**
Specify length of box: **6"**
Specify width of box or [Cube]: **26'**
Specify height of box: **9'**
Specify rotation angle of box about the Z axis or [Reference]: **0**

It is easiest to just copy the west wall to make the east wall:

1. Set SNAP to 6" and turn on and turn ORTHO on.
2. Use the COPY command 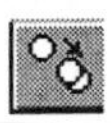 **and move to right side as illustrated (Fig. 12-1).**

Fig. 12-1 Copy the wall to make the east wall.

To draw the last wall, you can just point on the screen rather than entering coordinates:

1. Menu Bar: Draw > Surfaces.. > 3D Surfaces...
2. In the dialogue box, select the box by clicking on the picture icon and click on OK. Then respond to prompts:

Command: _ai_box
Specify corner point of box: **(Point A on Fig. 12-2 or 6 inches down from upper left corner of west wall)**
Specify length of box: **(point B on Fig. 12-2 or 6 inches down from upper right corner of east wall)**
Specify width of box or [Cube]: **6"**
Specify height of box: **9'**
Specify rotation angle of box about the Z axis or [Reference]: **0**

Fig. 12-2 Points to use for making north wall.

Your plan should look as in the illustration (Fig. 12-3).

Fig. 12-3 Plan with all three walls.

You will set up viewports so that you can see your room in 3D.

1. Menu Bar: View > Viewports > New Viewports...
2. In the Viewports window, highlight "Four: Equal" and press "OK."

The 4 viewports all have the same plan view. You will learn how to make different views for the viewports.

3DORBIT Command

The 3DORBIT command enables you to create views by dragging the crosshair. The icon is found on the standard toolbar.

1. Click on the upper right viewport to make it the current viewport.
2. Turn SNAP off by clicking on the status line until SNAP is forward.

Command: **3DORBIT**

1. Click on one of the small circles on the large circle and drag to rotate room.
2. Right click to see a shortcut menu, pan the view, right click, and click on "Exit" to exit this command.

The short cut menu also enables you to generate views such as "SE Isometric" or elevation views. You will do these later from the menu bar.

PLAN Command

The PLAN command generates a view parallel to the UCS. Since the UCS is WCS, it will generate a floor plan view as you had before using the 3D ORBIT command.

Command: **PLAN**
Enter an option [Current ucs/Ucs/World] <Current>: **(press ENTER)**

DVIEW Command

The DVIEW command also generates views. You will use it later to generate a perspective view.

Command: **DVIEW**
Select objects or <use DVIEWBLOCK>: **ALL**
Select objects or <use DVIEWBLOCK>: **(press ENTER)**
Enter option
[CAmera/TArget/Distance/POints/PAn/Zoom/TWist/CLip/Hide/Off/Undo]: **CA**
Specify camera location, or enter angle from XY plane, or [Toggle (angle in)] <90.0000>: **45**
Specify camera location, or enter angle in XY plane from X axis or [Toggle (angle from)] <90.0000>: **-45**
Enter option [CAmera/TArget/Distance/POints/PAn/Zoom/TWist/CLip/Hide/Off/Undo]: **H**
Enter option [CAmera/TArget/Distance/POints/PAn/Zoom/TWist/CLip/Hide/Off/Undo]: **(press ENTER)**

The display should look as in the illustration (Fig. 12-4).

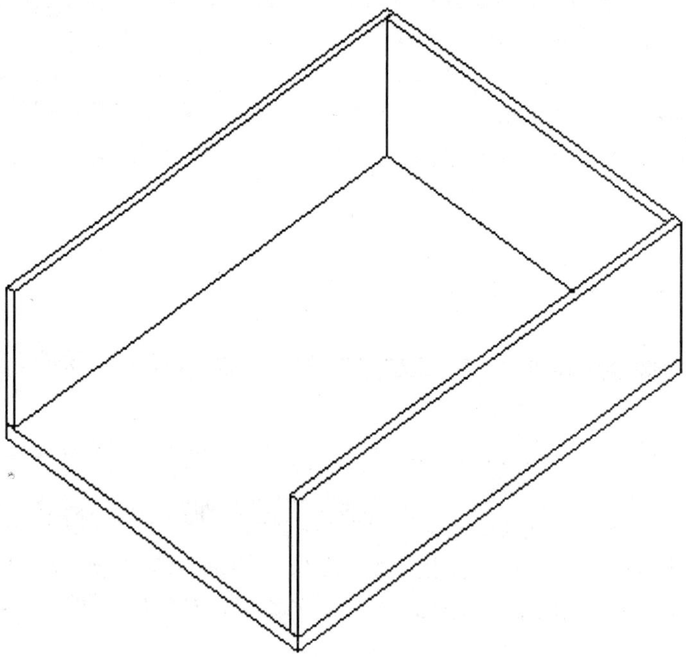

Fig. 12-4 Room with hidden lines removed.

1. Zoom to drawing extents to get full view.

You will save this view.

1. **Menu Bar: View > Named Views...**
2. **In View window, click on "New.."**
3. **Type "V1" in box next to "View name."**
4. **Click "OK" to exit New View window.**
5. **Click "OK" to exit View window.**

It is helpful to see more than one view so you will generate the right elevation in the lower left viewport:

1. **Click on the lower left viewpoint to make it current.**
2. **Menu Bar: View >3D Views > Right.**

The views should be similar to the illustration (Fig. 12-5.).

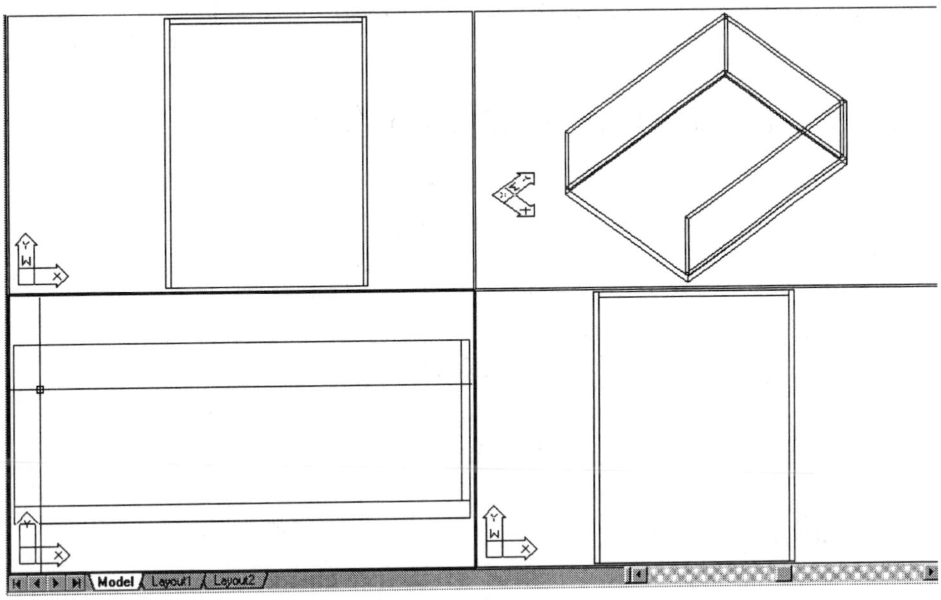

Fig. 12-5 Plan, V1, right elevation, and plan views in the four viewports.

Notice that the UCS is parallel to the right elevation. This means that you can draw in that elevation although they would be drawn outside of the room or about 1' from the limits origin. If you want to try dawing a box, do so and then erase the box before proceeding.

You will be drawing some columns for the room. The "3D Objects" dialogue box does not illustrate a cylinder so to draw cylinders, you use the CIRCLE command using the ELEV command to set the thickness or height for the column. Or, you could use cone and specify the same radius for top and bottom of the cone.

Command: **ELEV**
Specify new default elevation <1'-0">: **(press ENTER to keep elevation at 1')**
Specify new default thickness <0'-0">: **9'**

1. Change the current drawing layer to "columns."
2. Click on upper left viewport or a viewport showing the plan to make it current.

Command: **C**
CIRCLE Specify center point for circle or [3P/2P/Ttr (tan tan radius)]: **10',10'**
Specify radius of circle or [Diameter]: **6"**

You will use the ARRAY command to make four columns:

Command: **ARRAY**
Select objects: **L (or click on the circle)**
Select objects: **(press ENTER or right click)**
Enter the type of array [Rectangular/Polar] <R>: **R**
Enter the number of rows (---) <1>: **2**
Enter the number of columns (|||) <1> **2**
Enter the distance between rows or specify unit cell (---): **15'**
Specify the distance between columns (|||): **10'**

HIDE Command

The HIDE command removes hidden lines for a one-time view. Note: you do not use this for printing; there is a separate hide option on the plot window.

1. Click on the upper right viewport to make it current.

Command: **HIDE**

Next, you will make some 3D artwork for the north wall. You will align the UCS to the back or north wall. You do this by first rotating UCS around the X axis and then attaching the UCS origin to the lower left corner of the back wall. First, you must set the thickness back to 0 and set the elevation to 6" so that you can draw on the inside wall surface:

Command: **ELEV**
Specify new default elevation <0'-0">: **6"**
Specify new default thickness <9'-0">: **0**

1. In the upper right viewport, use the ZOOM command for a closer view (Fig. 12-6)

Command: **UCS**
Origin/ZAxis/3point/Entity/View/X/Y/Z/Prev/Restore/Save/Del/?/<World>: **X**
Rotation angle around X axis: **90**
Command: **UCS**
Origin/ZAxis/3point/Entity/View/X/Y/Z/Prev/Restore/Save/Del/?/<World>: **O**
Origin point <0,0,0>: **INT**
of **(point A)**
Command: **UCSICON**
ON/OFF/All/Noorigin/ORigin <ON>: **OR**

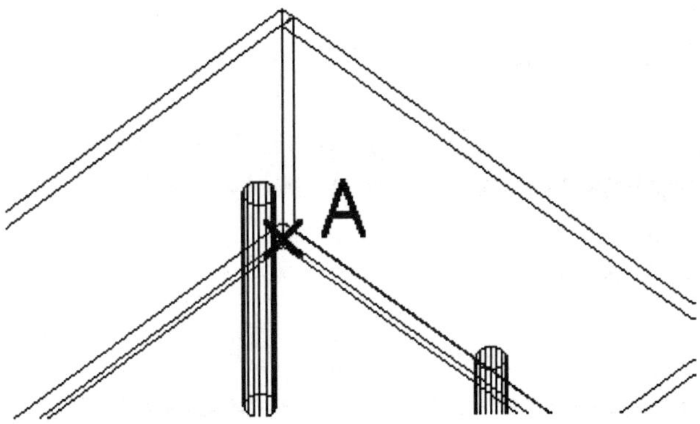

Fig. 12-6 Select the inside corner of the room.

The UCS should appear in the vertical direction or as illustrated (Fig. 12-7):

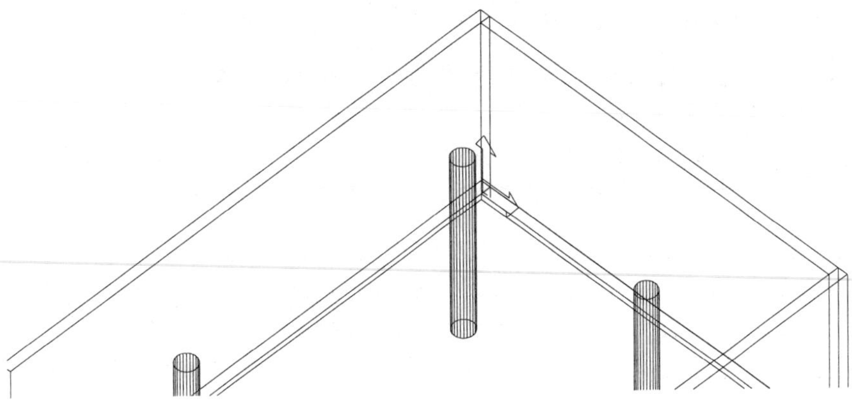

Fig. 12-7 UCS icon should be vertical.

You can also find the UCS command in the flyout UCS tool on the Standard Toolbar. Rotating on the X axis is the tool while you can place the origin by using the tool. This time, you will use the PLAN command to make an elevation view or view parallel to the UCS.

view parallel to the UCS.

Command: **PLAN**
Enter an option [Current ucs/Ucs/World] <Current>: **(press ENTER or right click)**

The upper right viewport should be an elevation now. You will restore the "V1" view in the lower right viewport so that you can see your artwork in 3D view.

1. **Click on the lower right viewport to make it the current viewport.**
2. **Make "3Dart" the current drawing layer.**

Command: **VIEW**

1. **In View window, highlight "V1" and click "Set Current" and "OK."**
2. **Click on the upper right viewport (elevation) to make it current.**

Now you will draw a dome:

1. **Menu Bar: Draw > Surfaces.. > 3D Surfaces...**
2. **When window appears, select the dome by clicking on the picture icon and click on "OK." Then respond to prompts:**

Specify center point of dome: **(Click on wall in elevation viewport)**
Specify radius of dome or [Diameter]: **6"**
Enter number of longitudinal segments for surface of dome <16>: **8**
Enter number of latitudinal segments for surface of dome <8>: **4**

1. **Make more artwork with the "3D Objects" feature.**

Your drawing may looks similar to the illustration (Fig. 12-8). Note that you can use the MOVE command to arrange your objects on the wall.

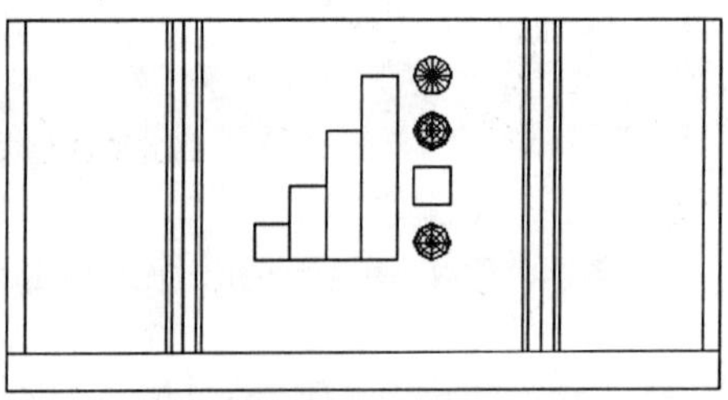

Fig.12-8 Elevation of artwork.

It is useful to save this viewport configuration for possible future use.

Command: **VPORTS**

1. Type "VP4" in box next to "New Name:" and Click "OK" to exit.

The viewport configuration with the same views in each viewport are saved.

Lastly, you will want to enter text to add to the "V1" view or view in the upper right viewport so you can print out this view at the end of this tutorial. In the illustration (Fig. 12-9), you can see that "Lobby Concept" was entered with AutoCAD text. You can use the UCS command with "View" option to align the UCS with your view or picture plane:

1. Make "text" the current layer.

Command: **UCS**
Enter an option [New/Move/orthoGraphic/Prev/Restore/Save/Del/Apply/?/World] <World>: **V**

1. Use DTEXT command pointing for the height or use 2 feet.
2. To print, make sure that the upper right viewport is current, and in the Plot Settings tab, use "Scaled to Fit" and check "Hide objects" on "Plot options."

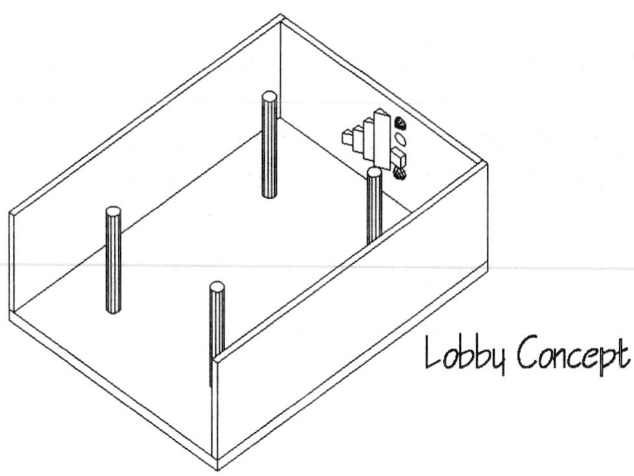

Fig. 12-9 Final drawing with hidden lines removed.

need to restore full views in the viewports.

1. Click on the upper right viewport and ZOOM all .

2. Click on the lower right viewport and ZOOM all .

TUTORIAL 13

3D FURNITURE

Commands Learned:	3DFACE	REVSURF
	TABSURF	RULESURF

In the previous tutorial, you learned how to use simple geometric forms to create an interior. At times, you will need to make more complex forms, particularly for furniture. For these, AutoCAD provides other commands which you will learn by making various pieces of furniture.

1. Open the file you created in the previous tutorial.
2. Make "furniture" the current drawing layer.
3. Freeze the "text" layer (be sure to freeze instead of just turn off).
4. Use the ELEV command and set the elevation to 1' keeping thickness at 0.

You will be drawing a reception counter for the lobby interior. First draw the shape pictured in the illustration and in the approximate position (Fig. 13-1). To draw the lines on an angle, rotate the UCS on a 45 degree angle (UCS - Z).

1. Set SNAP to 1".
2. Turn ORTHO on.
3. Click on the upper left viewport or the viewport with the floor plan to make it current. .
4. Using the LINE command, Draw the shape pictured with a 2' depth to the counter—the lines will probably overlap (Fig. 13-1). To draw line on 45 degree angle, rotate UCS 45 degrees on X axis.
5. Use the TRIM command to make all lines connect (Fig. 13-2).

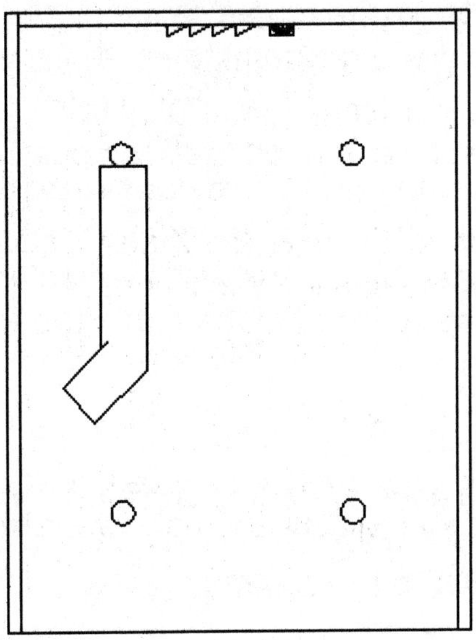

Fig. 13-1 Draw an angled desk.

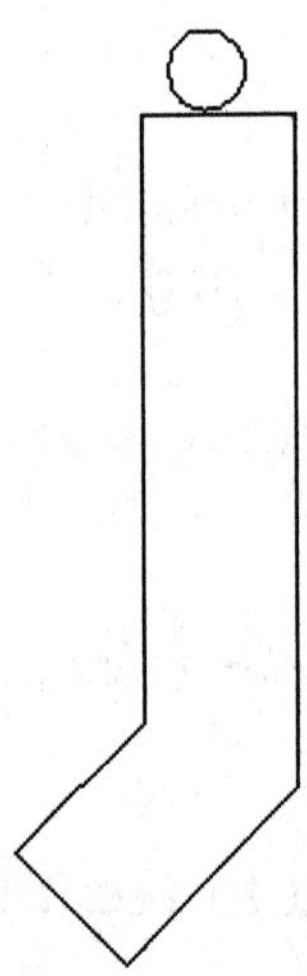

Fig. 13-2 Trim lines to make them connect precisely.

You will be drawing a copy of the square 18" off the ground. There are several ways to do this including changing the UCS, but since you know exactly where you want the copy, you can just enter the X,Y, Z coordinates. First:

1. Turn SNAP and ORTHO off.

2. Use the ZOOM command  to see just the reception counter.

Then use the COPY command:

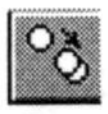

Command: **COPY**
Select objects: **(select all 6 lines)**
Select objects: **(press ENTER or right click)**
Specify base point or displacement, or [Multiple]: **0,0**
Specify second point of displacement or <use first point as displacement>: **0,0,3'**

The 3' was the Z coordinate or height while the X and Y stayed the same because you wanted the copy to stay in the same XY location.

3DFACE Command

The 3DFACE command allows you to make surfaces. You will use it to make the reception counter have faces. You will put another view in the lower right viewport so that points will be easier to select:

1. Click on the lower right viewport to make it current.

Command: **DVIEW**
Select objects or <use DVIEWBLOCK>: **all**
Select objects or <use DVIEWBLOCK>: **(press ENTER)**
Enter option
[CAmera/TArget/Distance/POints/PAn/Zoom/TWist/CLip/Hide/Off/Undo]: **CA**
Specify camera location, or enter angle from XY plane, or [Toggle (angle in)]
<45.0000>: **60**
Specify camera location, or enter angle in XY plane from X axis, or [Toggle
(angle from)] <45.00000>: **-30**
Enter option
[CAmera/TArget/Distance/POints/PAn/Zoom/TWist/CLip/Hide/Off/Undo]: **(press ENTER)**

1. Use the ZOOM command to see just the counter.

Refer to intersection points in sketch (Fig. 13-3).

1. **Menu Bar: Tools > Drafting Settings.**
2. **Click on Object Snap Tab.**
3. **Set object snap to "intersection," clear rest, and click "OK" to exit.**
4. **Click OSNAP on status line to turn on.**
5. **Turn ORTHO and SNAP off.**

Command: **3DFACE**
Specify first point or [Invisible]: **(point A)**
Specify second point or [Invisible]:**(point B)**
Specify third point or [Invisible] <exit>:**(point C)**
Specify fourth point or [Invisible] <create three-sided face>: **(point D)**
Specify third point or [Invisible] <exit>: **(point E)**
Specify fourth point or [Invisible] <create three-sided face>: **(point F)**
Specify third point or [Invisible] <exit>: **(point G)**
Specify fourth point or [Invisible] <create three-sided face>: **(point H)**
Specify third point or [Invisible] <exit>: **(point I)**
Specify fourth point or [Invisible] <create three-sided face>: **(point J)**
Specify third point or [Invisible] <exit>: **(point K)**
Specify fourth point or [Invisible] <create three-sided face>:**(point L)**
Specify third point or [Invisible] <exit>:**(point at intersection A again)**
Specify fourth point or [Invisible] <create three-sided face>: **(point B again)**
Specify third point or [Invisible] <exit>: **(press ENTER or right click)**

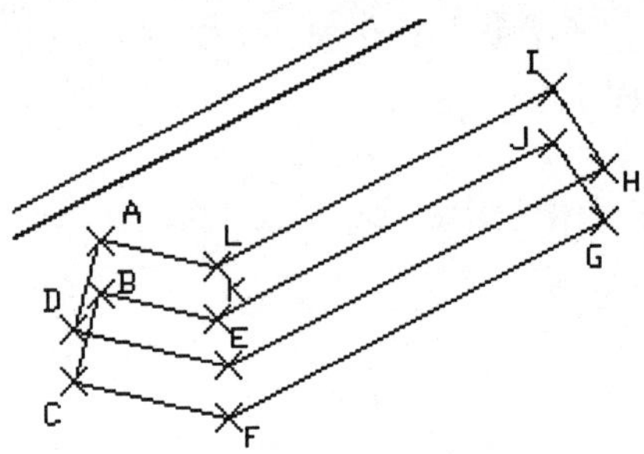

Fig. 13-3 Using intersection object snap, click on points as illustrated.

Command: **HIDE**

All the sides of the counter should appear closed (Fig. 13-4).

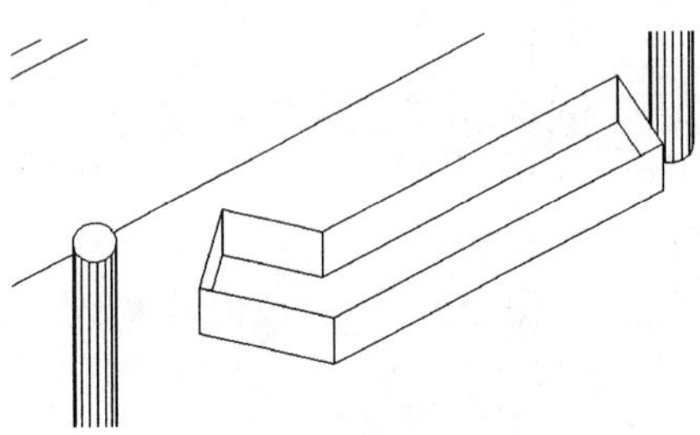

Fig. 13-4. Counter without top.

You will use the 3DFACE command again to make the top referring to points in the illustration (Fig. 13-5). But first you will bring up the "Surfaces" toolbar.

1. **On menu bar, click on "View."**
2. **Click on "Toolbars..."**
3. **Check "Surfaces"**
4. **Click on "Close" and place the toolbar where convenient to use.**

3DFACE
Specify first point or [Invisible]: **(point A)**
Specify second point or [Invisible]:**(point B)**
Specify third point or [Invisible] <exit>:**(point C)**
Specify fourth point or [Invisible] <create three-sided face>: **(point D)**
Specify third point or [Invisible] <exit>: **(point E**
Specify fourth point or [Invisible] <create three-sided face>: **(point F)**
Specify third point or [Invisible] <exit>: **(press ENTER or right click)**

Command: **HIDE**

The HIDE command can also be found under "View" on the menu bar. If the top does not appear to be closed, undo and try again.

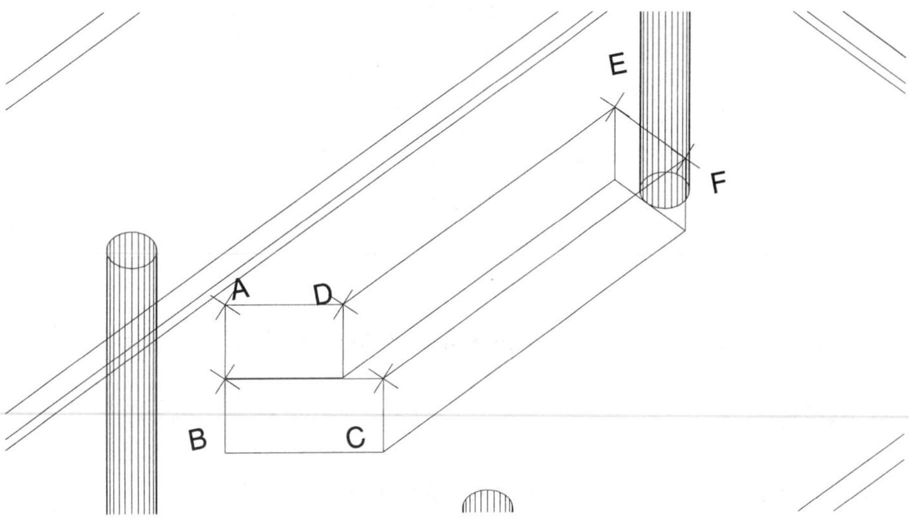

Fig. 13-5 Use points for the 3DFACE command.

3D Meshes

Besides the 3D objects, AutoCAD has a number of 3D commands that can help you draw any object. AutoCAD provides a way to put a surface on any shape object. It does so by creating meshes, much like draping a fabric over a shape. The commands that create meshes are: RULESURF, TABSURF, REVSURF, and EDGESURF. AutoCAD also provides system variables (SURFTAB1 and SURFTAB2) to control the density of the mesh.

1. Click on the upper left viewport or the viewport with the floor plan to make it current.

2. Draw a 12" radius circle with the CIRCLE command in the middle of the room.

3. Turn off OSNAP by clicking on the status line.

With the 12" radius circle you just drew, you will make a cone shape or a plant stand; then you will draw a second smaller circle at 30" height. You will also be using the .xy filter feature of AutoCAD which allows you to extract the x and y coordinates of an entity to draw another entity. The command sequence is:

CIRCLE
Specify center point for circle or [3P/2P/Ttr (tan tan radius)]: **.xy**
of **cen (or click on)**
of **(point to 12" radius circle)**
(need Z): **30"**
Specify radius of circle or [Diameter] <1'-0">: **6"**

A second smaller circle should be drawn 30" above the first circle.Check in the other viewport to make sure you have drawn the circle directly over the bottom circle. If not, redo making sure you use "center" object snap to AutoCAD's second prompt. Now you will complete the cone with the RULESURF command.

RULESURF Command

The RULESURF command will apply a surface connecting any two known entities, in this case, the two circles. You could do this also with two arcs of different shapes, two lines, a point and a circle, a point and a line, a triangle and a point (to make a pyramid), and so on.

Command: **RULESURF**
Current wire frame density: SURFTAB1=6
Select first defining curve: **(point to lower circle)**
Select second defining curve: **(point to upper circle)**

The cone shape is not very smooth (Fig. 13-6).

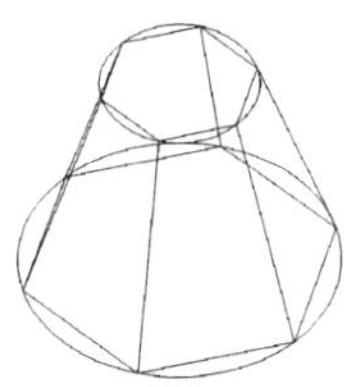

Fig. 13-6 Initial work with RULESURF.

1. ERASE the rulesurf by pointing on the ruled surface.

You need to set SURFTAB1, a system variable controlling the vertical mesh, to a higher value. The sequence is:

Command: **SURFTAB1**
Enter new value for SURFTAB1 <6>: **20**

1. Redo the RULESURF on the two circles.
2. Use HIDE command to remove hidden lines.

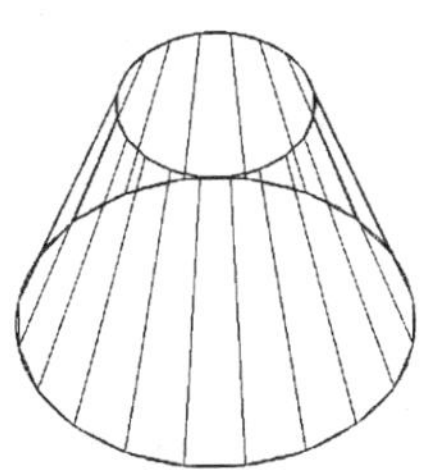

Fig. 13-7 A smoother cone by setting mesh variable to a higher value.

The easier way of making this cone shape was to use cone on "3Dobjects" setting

the radii for the bottom and top to be 12" and 6" respectively.
Next, you will draw a curved room screen 5' in height. To do so draw a polyline as in the illustration with the approximate measurements shown. You will need to return to plan view (using PLAN command) or draw the polyline in the upper left viewport.

1. Click on the floor plan viewport to make it current.
2. Turn OSNAP off.
3. Use the PLINE comand to make a line similar to illustration (Fig. 13-8).
4. Use PEDIT with "F" otion to make it curved (PEDIT - F) as in illustration (Fig. 13-9).

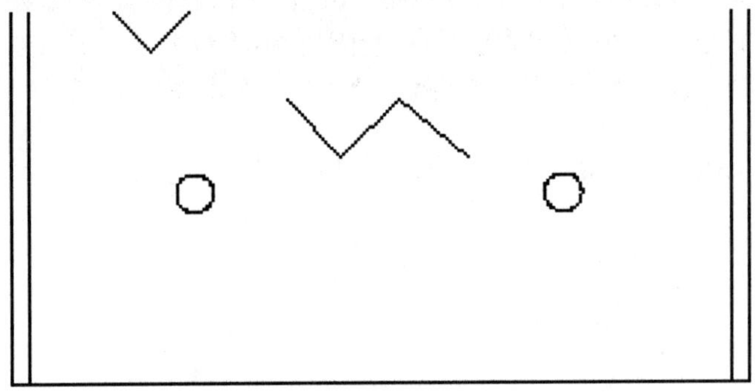

Fig. 13-8 Draw a polyline.

Fig. 13-9 Use "Fit" with PEDIT command to make a curve.

You will set a new UCS so that you can draw a line in the Z direction:

Command: **UCS**
Origin/ZAxis/3point/Entity/View/X/Y/Z/Prev/Restore/Save/Del/?/<World>: **X**
Rotation angle about the X-axis <0,0>: **90**

1. In the lower right viewport (containing a 3D view), use ZOOM to make sure you can see the curved line.

To draw a line which will function as a direction vector for the TABSURF, you need to enter the coordinates as below:

Command: **L**
LINE Specify first point: **endp**
of **(point to curve at one end)**
Specify next point or [Undo]: **@0,5'**
Specify next point or [Undo]: **(press ENTER or right click)**

By typing "@," you are telling AutoCAD that you want the line to start at the same X and Y location , 0 units in the X direction, and 5 feet in the Y direction. Because Z is 0, you did not have to enter this. Your line should look as illustrated (Fig. 13-10).

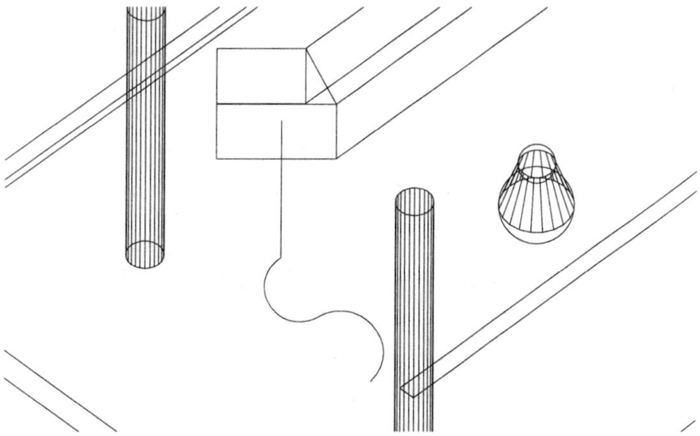

Fig. 13-10 The line should appear vertical or in the Z direction.

1. Return the UCS to the World Coordinate System (UCS - press ENTER) in the viewport you drew the line.

TABSURF Command

Now that you have set it up, the TABSURF command easily draws the curved screen:

Command: **TABSURF**
Select path curve: **(point to polyline)**
Select direction vector: **(point to 5' line)**

You did not have to draw the direction vector on the polyline itself. If the TABSURF went below the floor line, undo the TABSURF and then redo selecting a different (usually lower) point on the direction vector. Your screen should appear as illustrated (Fig. 13-11).

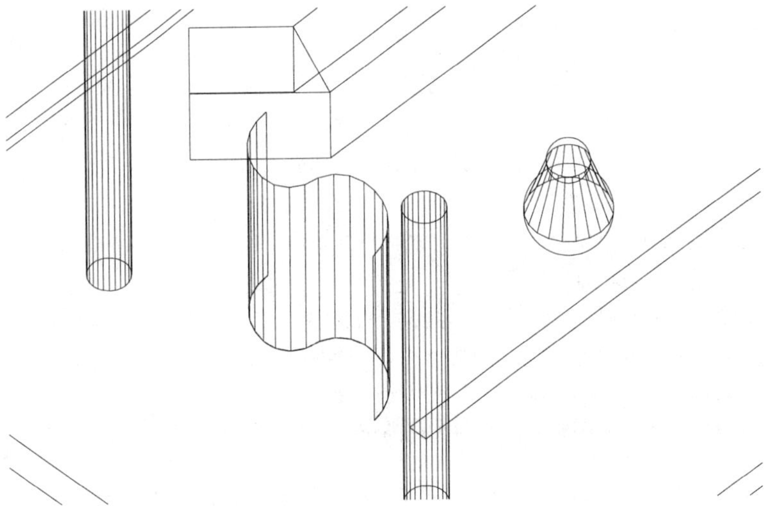

Fig. 13-11 The finished screen using TABSURF.

Finally, you will use the REVSURF command by making a round or tulip-shaped base for a table. In order to do this, you will need to draw two lines in elevation. You will also need to set the "Origin" on the lower corner of the reception counter or box to set an imaginary wall parallel to the box to draw the two lines. You will also move the UCSICON to that corner to see where the elevation is (Fig. 13-12):

1. Click on the lower right viewport to make it current.
2. If UCS is not WCS and the view not a 3D view, set UCS to WCS (UCS - press ENTER) and restore the V1 view.

Command: **UCS**

Current ucs name: *WORLD*
Enter an option
[New/Move/orthoGraphic/Prev/Restore/Save/Del/Apply/?/World] <World>: **X**
Specify rotation angle about X axis <90>: **90**
Command: **(press ENTER)**
UCS Current ucs name: *NO NAME*
Enter an option
[New/Move/orthoGraphic/Prev/Restore/Save/Del/Apply/?/World] <World>: **O**
Specify new origin point <0,0,0>: **INT**
of **(point A)**
Command: **PLAN**
Enter an option [Current ucs/Ucs/World] <Current>: **(press ENTER)**

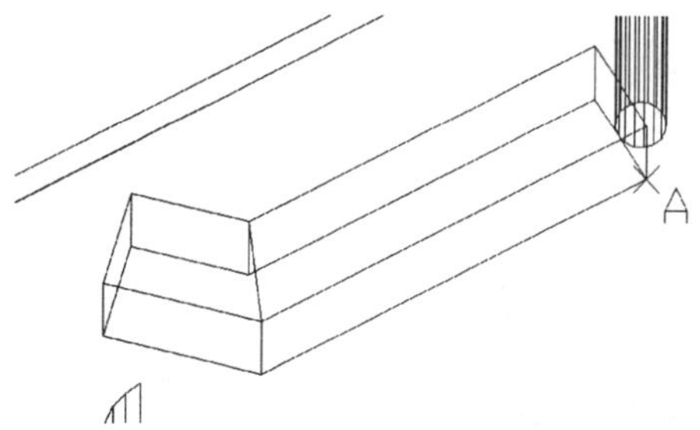

Fig. 13-12 Point for UCS origin.

1. Use PLINE command to draw line as in the illustration (Fig. 13-13) making the height about 30").
2. Edit the polyline using PEDIT and make it curve using Fit option.
3. Click on the upper right viewport to make it current.
4. ZOOM if necessary to see the two lines you just drew.
5. Set the SURFTAB2 system variable to 20 (SURFTAB2) .

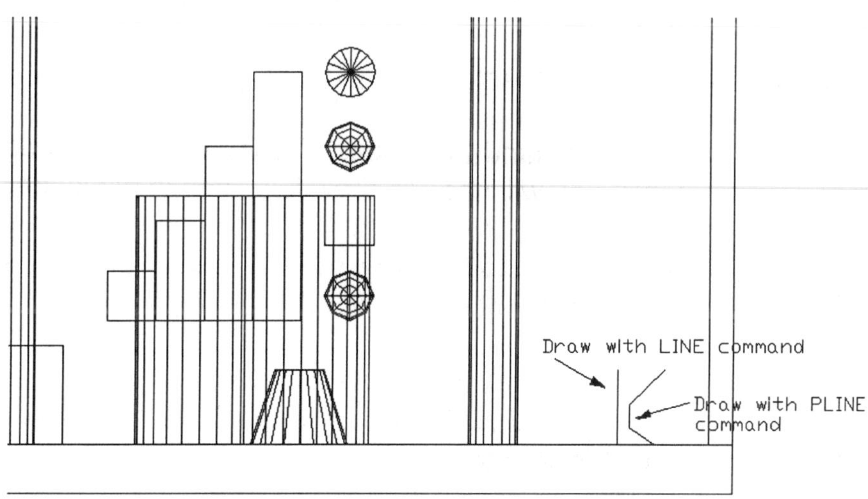

Fig. 13-13 Draw polylines.

REVSURF Command

The REFSURF command will revolve a mesh around an axis. PAN or ZOOM if necessary to view the polyline and line you just drew. The vertical line will be used as a direction vector for the REVSURF command.

1. **Click in the lower left viewport.**
2. **Set UCS to WCS (UCS - press ENTER).**
3. **Zoom to see the two lines you just drew.**

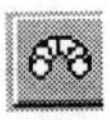

REVSURF
Current wire frame density: SURFTAB1=20 SURFTAB2=20
Select object to revolve: **(point to curved polyline)**
Select object that defines the axis of revolution: **(point to vertical line)**
Specify start angle <0>: **(press ENTER or right click)**
Specify included angle (+=ccw, -=cw) <360>: **(press ENTER or right click)**

The table base should look as in the illustration (Fig. 13-14).

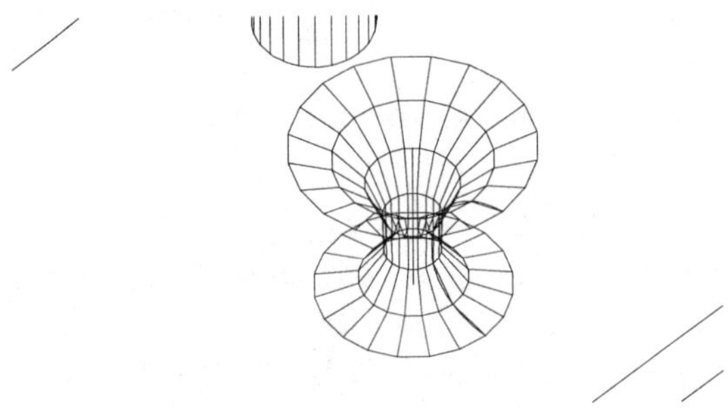

Fig. 13-14 Tulip shaped table base done with REVSURF command.

If you really understand how to work with AutoCAD 3D, you will be able to draw a 2" thick, 36" diameter table top for the table base without referring to the instructions below:

Command: **ELEV**
Specify new default elevation <0'-0">: **(press ENTER)**
Specify new default thickness <0'-0">: **2"**
Command: **C**
CIRCLE
Specify center point for circle or [3P/2P/Ttr (tan tan radius)]: **end**
of **(click on top end of vertical line in tulip base)**
Specify radius of circle or [Diameter] <1'-0">: **18"**

Finish the Interior

You will want to finish the interior lobby by adding more 3D furniture. But first you will turn the elevation back to 1' and thickness to 0 so that all objects will be on the floor.

Command: **ELEV**
Specify new default elevation <0'-0">: **1'**
Specify new default thickness <0'-0">: **0**

The chair was made in two parts using the "box" of "3D Objects." The seat height is 18" with a 30" height back. Also, you can arrange the furniture by using the MOVE command in the floor plan viewport (upper left) just as you do in 2D drafting. The corner round table was made with the CIRCLE command with thickness at 18."

This completes the 3-D furniture which you will use for your 3-D room (Fig. 13-15).

The room can be printed out by clicking on a viewport with a 3D view.

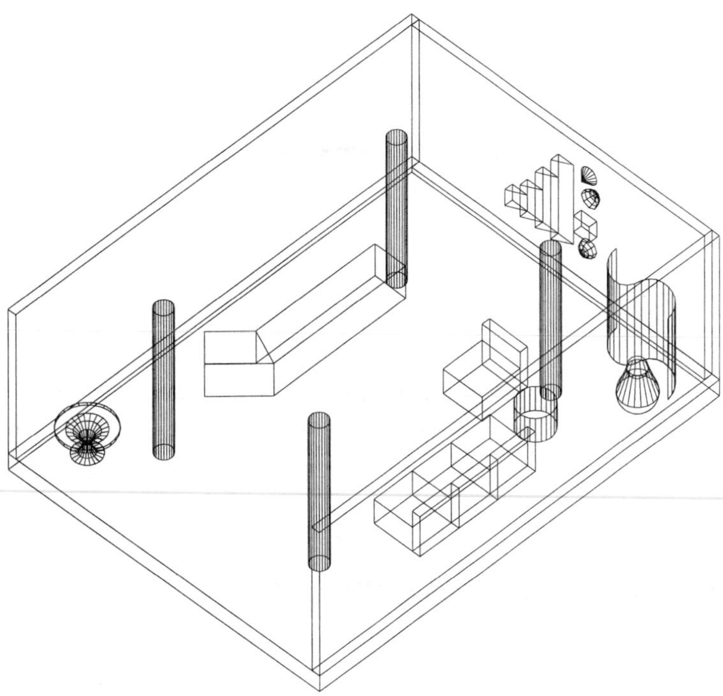

Fig. 13-15 Final room with added furniture.

TUTORIAL 14

PERSPECTIVE INTERIOR

Command Learned: EDGESURF

1. Open the file you used in the previous tutorial.

Your floor plan probably appears similar to the illustration (Fig. 14-1).

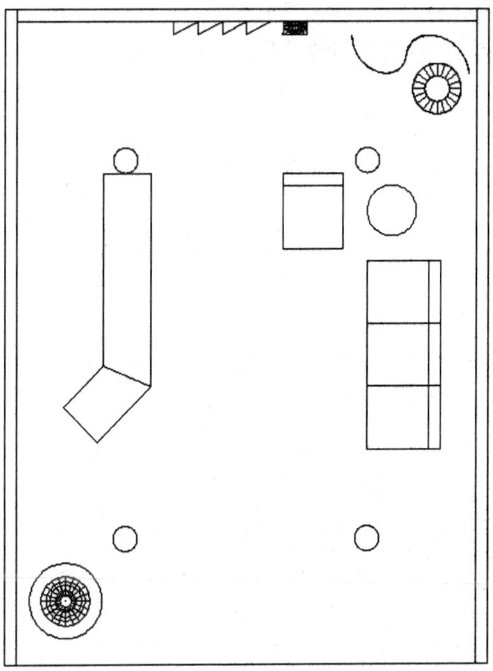

Fig. 14-1 Floor plan of 3D Room.

Set elevation and thickness back to 0:

Command: **ELEV**
Specify new default elevation <1'-0">: **0**
Specify new default thickness <0'-0">: **0**

In the following sequence, you will align the UCS to the wall behind the reception counter by first rotating the UCS around X axis, then Y. Refer to the illustration for Point A for placing the origin of the UCS (Fig. 14-2).

1. Click on upper right viewport (and restore V1 view if necessary).
2. Zoom to the lower left corner of the room (Fig. 14-2).

Command: **UCS**
Current ucs name: *WORLD*
Enter an option [New/Move/orthoGraphic/Prev/Restore/Save/Del/Apply/?/World] <World>: **X**
Specify rotation angle about X axis <90>: **90**
Command: **(press ENTER)**
UCS
Current ucs name: *NO NAME*
Enter an option [New/Move/orthoGraphic/Prev/Restore/Save/Del/Apply/?/World] <World>: **Y**
Specify rotation angle about Y axis <90>: **90**
Command: **(press ENTER)**
UCS
Current ucs name: *NO NAME*
Enter an option [New/Move/orthoGraphic/Prev/Restore/Save/Del/Apply/?/World] <World>: **O**
Specify new origin point <0,0,0>: **INT (or click on** 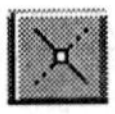 **icon)**
of **(Point A)**
Command: **UCSICON**
Enter an option [ON/OFF/All/Noorigin/ORigin] <ON>: **OR**

The UCS icon should be in the lower left corner of the inside wall and parallel to the wall.

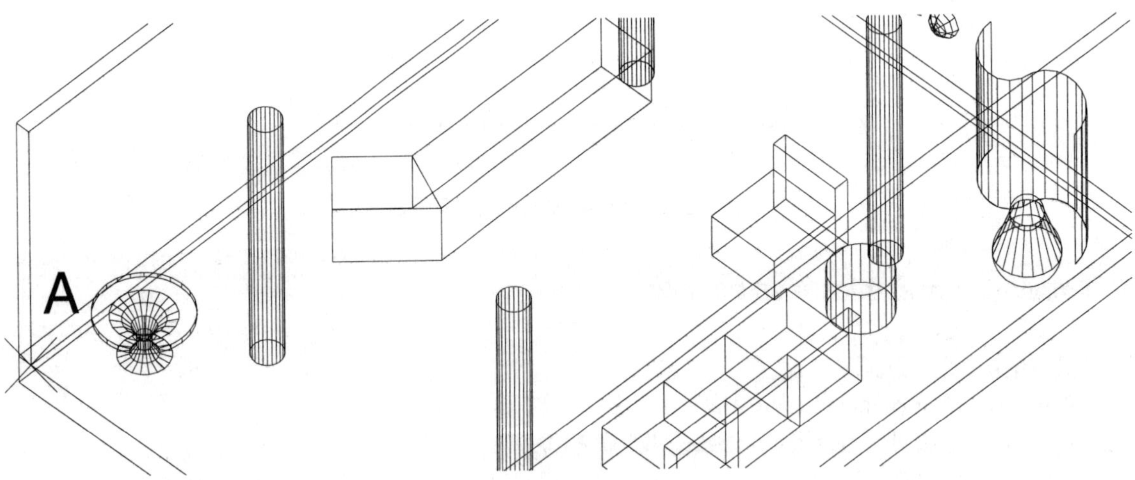

Fig. 14-2 Align UCS to wall.

1. Make a new layer "artwork"and make it the current drawing layer.

Because the UCS is aligned on the wall plane, use the PLAN command to make an elevation of the wall in the upper right viewport.

Command: **PLAN**
<Current UCS>/Ucs/World: **(press ENTER or right click)**

You will be drawing a mesh artwork on the wall using the EDGESURF command. This command requires 4 edges to make a mesh so you will first make 4 polylines for the edges.

1. Use PLINE command to draw the 4 separate polylines as illustrated (Fig. 14-3) making sure each polyline is connected to the other one (by using endpoint object snap) approximately at the points A, B, C, and D.

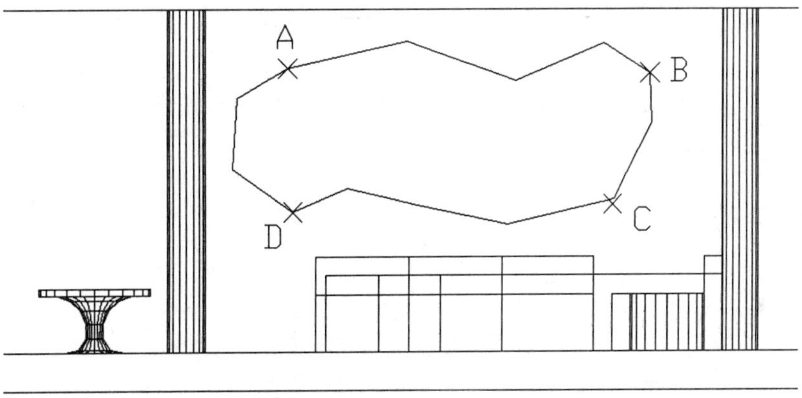

Fig. 14-3 Connect lines precisely to each other by using object snap.

2. Using the PEDIT command, edit each polyline using the "Fit curve" option to make the following shape (Fig. 14-4).

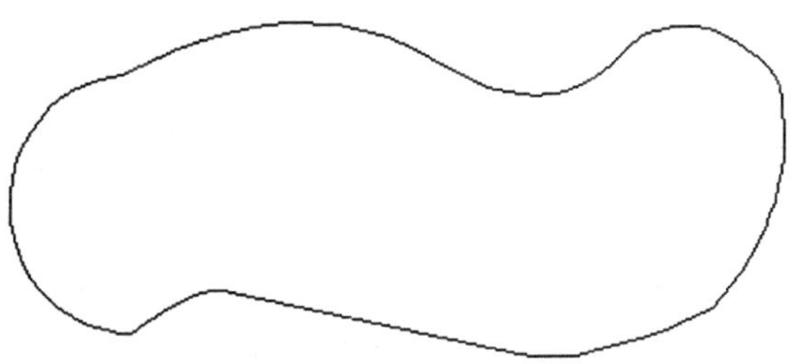

Fig. 14-4 Curved polylines.

EDGESURF Command

The EDGESURF command is used to make a mesh from four edges. These could be polylines, arcs, etc. In the next sequence, you will set SURFTAB1 and SURFTAB2 variables to 30. Then you will use EDGESURF command to make a 3-D mesh.

Command: **SURFTAB1**
Enter new value for SURFTAB1 <20>: **30**
Command: **SURFTAB2**
Enter new value for SURFTAB2 <6>: **20**

Command: **EDGESURF**
Current wire frame density: SURFTAB1=30 SURFTAB2=20
Select object 1 for surface edge: **(point to the first polyline you drew)**
Select object 2 for surface edge: **(point to the second)**
Select object 3 for surface edge: **(point to the third)**
Select object 4 for surface edge: **(point to the last polyline)**

Your mesh or artwork should appear similar to the illustration (Fig. 14-5). The mesh will be less dense if you had set SURFTAB1 and SURFTAB2 to a lower value.

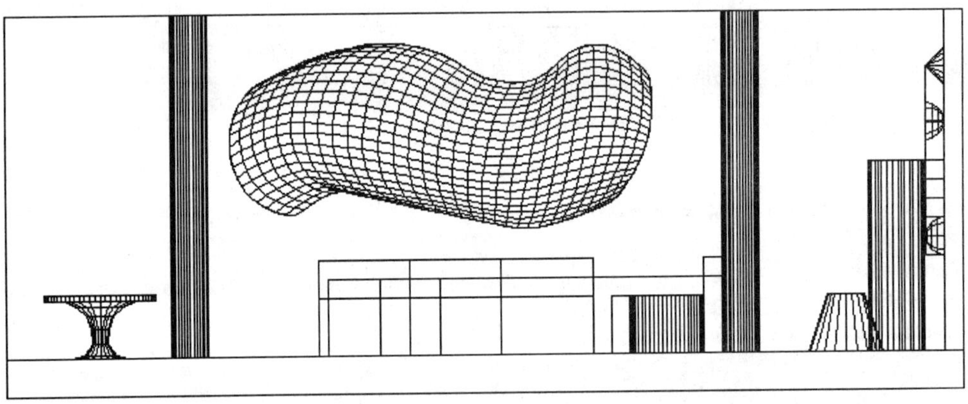

Fig. 14-5 Artwork created with EDGESURF.

The artwork is flat against the wall. If you had wanted to put some depth, you would have needed to draw in different UCS planes to create a 3D mesh.

Other Mesh and 3D Edit Commands

A mesh has horizontal lines (N direction) and vertical lines (M direction). If you know precisely where you would want each vertex or intersection of the M and N mesh, you could use the 3DMESH command to create the size of the mesh (N x M) and the coordinate locations of each vertex. This is covered in the AutoCAD User's Guide.

There are a few 3D edit commands as well. The 3DARRAY command, found under "Modify" and then "3D Operation..." on the menu bar, is similar to the ARRAY command. The rectangular option has an additional level option to add to the rows and columns to make a 3D array. The polar option asks for a second point on the axis of the rotation in addition to the center point to achieve a circular 3D array.

The 3DMIRROR command, also found under "Modify" and then "3D Operation..." on the menu bar, is similar to the MIRROR command but you specify three points to define the mirroring plane.

Next, you make a perspective view of your lobby.

1. Click on the upper left or plan viewport to make it current.

2. ZOOM to see all of floor plan.

Making A Perspective View

You will be using the DVIEW command to make a dynamic or perspective view of the interior. For points, look on the illustration below (Fig. 14-6). You first enter the target point (where you are looking) and then the camera point (where you are viewing from). You need to use the xy filter so you will be prompted for a height for these points. Finally, you will enter a distance to give the view perspective view.

Command: **DVIEW**
Select objects or <use DVIEWBLOCK>: **ALL**
Select objects or <use DVIEWBLOCK>: **(press ENTER)**
Enter option
[CAmera/TArget/Distance/POints/PAn/Zoom/TWist/CLip/Hide/Off/Undo]: **PO**
Specify target point <17'-1 3/8", 18'-0", 5'-4 3/16">: **.XY**
of **(point A)**
(need Z): **3'**
Specify camera point <17'-1 3/8", 18'-0", 5'-5 3/16">: **.XY**
of **(point B)**
(need Z): **5'**
Enter option
[CAmera/TArget/Distance/POints/PAn/Zoom/TWist/CLip/Hide/Off/Undo]: **D**
Specify new camera-target distance <24'-0">: **55'**
Enter option
[CAmera/TArget/Distance/POints/PAn/Zoom/TWist/CLip/Hide/Off/Undo]: **PA**
Specify displacement base point: **(click on view)**
Specify second point: **(move to get whole view in viewport)**
Enter option [CAmera/TArget/Distance/POints/PAn/Zoom/TWist/CLip/Hide/Off/Undo]: **(press ENTER)**

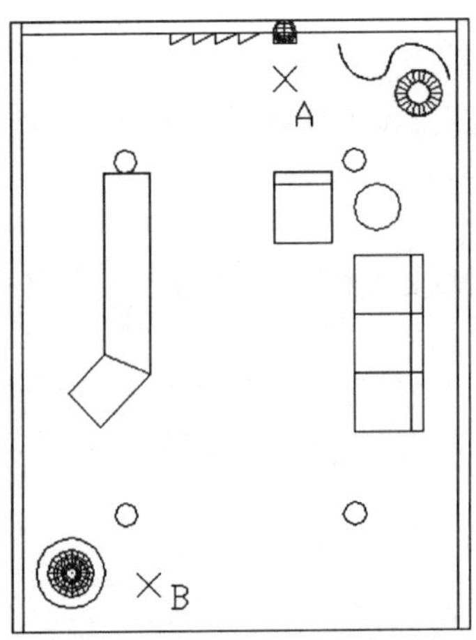

Fig. 14-6 Points to enter for creating a perspective view.

Although you could have used the bar chart to view a perspective dynamically, you often will know a good distance to specify. In this case, 55 feet should give you a good view (Fig. 14-7). At 70 feet you are farther away from the room while a distance of 20 feet puts you inside the room. Notice that a small perspective icon replaces the UCS icon in the lower left corner. This lets you know you are in perspective view. You cannot edit or use most commands in perspective view.

Fig. 14-7 Final perspective view.

1. **Save this view and name it "P1" by using the VIEW command.**
2. **Repeat the DVIEW command and generate other perspective views.**
3. **Save at least one more perspective view with the VIEW command.**

To plot these perspectives on a plotter, you can specify "view" as the area to plot and then AutoCAD will prompt you for the name of the view to plot. Or, just make the perspective view the current viewport and plot using "display" and "fit" for scale. Be sure to check "Hide objects" under Plot options to remove the hidden lines.

TUTORIAL 15

RENDERING

Commands Learned:	SHADE	SCENE
	RENDER	RMAT
	BACKGROUND	SETUV
	LIGHTS	

Rendering a three-dimensional wireframe drawing enables you to see how an interior might appear. This tutorial will show you how to use basic and photorealistic rendering. You will add lights, views, and materials to the interior you created in the previous 3D tutorials. You will also save your renderings to an image file which you can display later or print out on a color printer.

1. Open the drawing file you created in Tutorials 12-14.
2. Make the perspective drawing the active or current viewport.
3. Menu Bar: View >Viewports > 1 Viewport.

In the previous tutorials, you created your model by covering all the geometry with meshes by using 3D objects or the 3D mesh commands. In short, all the objects have surfaces which is required for rendering. You also did not overlap geometry or have two faces or polygons occupy the same space. If you did, you might get ambiguous or incorrect results with rendering. So, it is good practice not to overlap geometry.

RENDER Command

The RENDER command produces images that have shading, shadows, and assigned materials and thus can make a highly realistic image of your interior model.

Command: **RENDER**

1. In the Render dialogue box, make sure that "Smooth shade" is checked.
2. Click the "Render" button.

In a few seconds, a rendering of the perspective appears with the colors associated with each layer shown (Fig. 15-1).

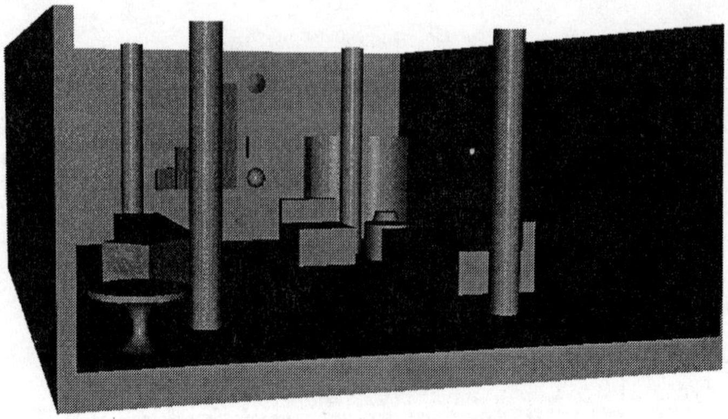

Fig. 15-1 Rendered image.

1. Repeat the RENDER command and remove check from "Smooth shade."

The cylinders are not smooth which results in a faster rendering. Another way to do a faster rendering on a complex image is to either check the "Query for Selections" to select one or two objects for rendering. Next, you will render part of the perspective by clicking on "Crop window" in the Render dialogue window.

1. Repeat the RENDER command.
2. Check "Smooth shade."
3. Check "Crop window."

Pick crop window to render: **(make a small window to select small area of room)**

The result may be similar to illustration (Fig. 15-2).

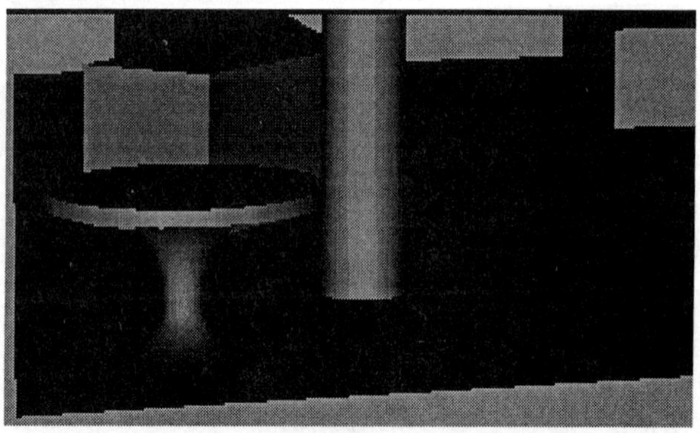

Fig. 15-2 Partial rendering by cropping window.

Rendering the Background

AutoCAD uses the color of the graphics window as a default background. You have the option, though, of selecting a solid color, a gradient, or an image for the background. You can do this by clicking on the background icon on the Render toolbar or through the Render dialogue window. You will change the background to a gradient:

1. Click on the Render icon.
2. Remove the check from the "Crop Window."
3. Click on the "Background..." button.
4. Click on "Gradient" button.
5. Click on the top box which is colored red.
6. Using the RGB (Red-Green-Blue) slider, change the color.
7. Change the colors for the middle and bottom box also.
8. Click on "OK" to exit the background window.
9. Click on "Render" to finish the rendering.

You should see a gradient of three colors. You will restore the AutoCAD background:

1. Repeat the RENDER command.
2. Click on the "Background..." button.
3. Click on "Solid" button.
4. make sure "AutoCAD background" is checked.
5. Click on "OK" to exit the background window.
6. Click on "Render" to finish the rendering.

LIGHT Command

Lighting can add realism to a rendering by modeling three-dimensional surfaces. AutoCAD adds a default light which gives an overall appearance of lighting. However, you will want to use other types of lights which can create more realistic interiors:

Ambient light. Provides a constant level of illumination similar to general lighting; has no direction. The level should be kept low or else the room will appear too bright and flat. Ambient lighting is useful as a fill light.

Point Light. Radiates light in all directions from its point location with its intensity diminishing the farther away from the point source (Fig. 15-3). The point light simulates an overhead general light well. Typically, you will use these lights at a low intensity to give general illumination.

Spotlights. Similar to track lighting, a spotlight makes a cone of light that can be angled in any direction depending on where you place the target and light source. The spotlight has a hotspot which is the brightest illumination and a falloff where the light falls off from the hotspot to no illumination at the edge of the cone (Fig. 15-3). The intensity of the light also diminishes over distance similar to the point light. Spotlights increase the rendering time but are worth it because they add strong contrast and drama to interiors.

Distant Lights. A distant light is most useful to simulate sunlight. The Sun Angle Calculator can determine the sun angle for different locations, time of day or dates. Thus, the distant light is useful for solar or daylighting studies.

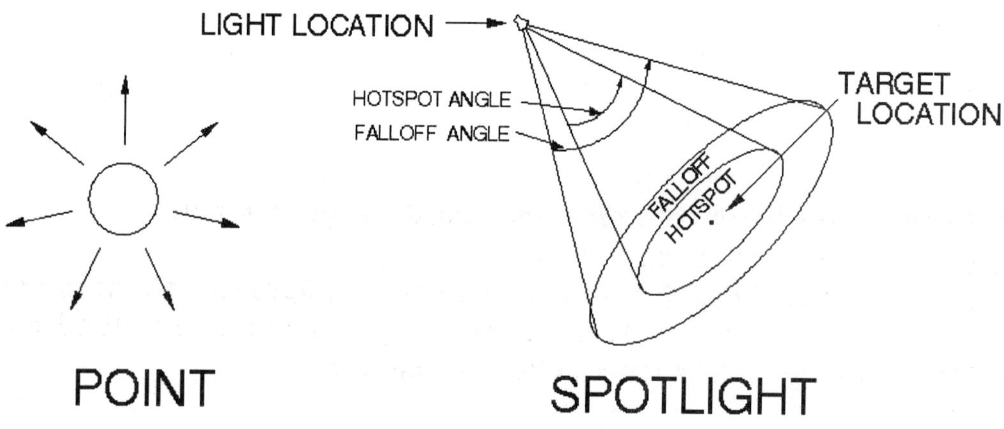

Fig. 15-3 Point emits light in all directions while spotlight emits directional light.

Lights can have different colors to simulate incandescent or other types of lights. This is done in the Lights window by changing the percentage of red, green, or blue (RGB) colors which are the primary colors of light.

You will need both plan and elevation views to place lights. Just for practice and to remember the name of the command, you will make a viewport configuration from the command prompt line:

Command: **-VPORTS**
Specify corner of viewport or
[ON/OFF/Fit/Hideplot/Lock/Object/Polygonal/Restore/2/3/4] <Fit>: **3**
Enter viewport arrangement
[Horizontal/Vertical/Above/Below/Left/Right] <Right>: **R**
Specify first corner or [Fit] <Fit>: **(press ENTER)**

1. **Click on the upper left viewport.**
2. **Menu Bar: View > 3D Views > Top.**
3. **Menu Bar: View > Shade > 2D Wireframe.**

You returned the image to a line or wireframe view.

1. **Click on the lower left viewport.**
2. **Menu Bar: View > 3D Views > Right.**
3. **Menu Bar: View > Shade > 2D Wireframe.**

You should have the floor plan in the upper left viewport, the right elevation in the lower left viewport, and the perspecive view V1 in the right viewport. You will place lights in the plan viewport using the .xy filter so AutoCAD will prompt for the height. Placing lights in the camera or perspective viewport should not be done because you cannot be sure of their location.

Command: **LIGHT**

1. **Click on the "New" button in the Lights dialogue menu.**
2. **Enter "point" for the light name.**
3. **Use the RGB slider to adjust the color of the light to a slight reddish color by making red at full saturation and green and blue a little less.**
4. **Under the Position section, click on "Modify..."**

Enter light location <current>: **.xy**
of **(point near the front of the room)**
(need Z): **8'**

1. **Click "OK" twice to exit Lights window.**
2. **Render the perspective viewport.**

The result should appear as illustrated (Fig. 15-4).

Fig. 15-4 Rendering with one point light.

The light appears very bright so you will modify its intensity.

Command: **LIGHT**

1. **Click on the "Modify" button in the Lights window.**
2. **Type 245 in box next to "Intensity."**
3. **Click "OK" to close Modify Point Light window.**
4. **Click "OK" to close Lights window.**
5. **Render the perspective viewport.**

You can see that the light drops off at the back of the room. You will put a spotlight to light the artwork on the back wall using .xy filters so you will be prompted for the height of the light source again:

Command: **LIGHT**

1. **Next to "New button, click on the down arrow and change from Point to Spotlight.**
2. **Click on the "New" button.**
3. **Enter "spot" for the light name.**
4. **Under the Position section, click on "Modify..."**

Enter light target <current> **.xy**

of **(point near back wall on art work in the plan viewport)**
(need Z): **4'**
Enter light location <current>: **.xy**
of **(point about 3-4' back from the back wall in the plan)**
(need Z): **8'**

5. Click "OK" twice to exit Lights window.
6. Render the perspective viewport.

Fig. 15-5 Rendering with spotlight on the back wall.

The spotlight misses the top part of the objects (Fig. 15-3). You can adjust the falloff to correct that:

Command: **LIGHT**

1. Highlight the spotlight and click on "Modify.."
2. Change the falloff to 90.
3. Render the perspective viewport.

Another type of AutoCAD light is the distant light. You might use this to simulate the sun coming through a window. The Sun Angle calculator will calculate the sun angle for different locations.

1. Next to "New" button, click on the down arrow and change from Spotlight to Distant.
2. Click on the "New" button.
3. Enter "distant" for the light name.
4. Render the perspective viewport.

SCENE Command

The SCENE command enables you to create renderings with different lighting. It is useful for lighting studies or if you have multiple views you want to render.

1. Click on the Scenes icon.
2. Click on "New..."
3. Type "distant" for scene name.
4. Select only Distant for Lights and then 'OK" to close window.
5. Click on OK to close Scenes window.
6. Render the perspective clicking on "distant" scene to render.
5. Click on the Scenes icon.
6. Click on "New..."
7. Type "ptspot" for scene name.
8. Click on point and hold down mouse button to also select spotlight for Lights and then click "OK" to close.
9. Click on "OK" to close Scenes window.
10. Render the perspective using "ptspot" scene to render.

This scene should have rendered differently since different lights were selected for each scene.

RMAT Command

You can create materials to make your interior look more realistic.

1. Click on the Materials icon.
2. Choose "New" in the Materials dialogue box.
3. Enter the name "fabric" in the material name box in the New Standard Materials dialog box.

All materials have properties which you can define in this dialogue box by clicking each property and adjusting its associated values.

1. For your fabric, try clicking on each of the properties listed below.
2. Render the preview to see the effect of changing each property.

Color/Pattern: If the "By ACI" box is unchecked, you can change the basic color of the material which can be defined by adjusting the RGB slider for hue or adjusting the value slider which controls lightness or darkness.

Ambient: color reflected from ambient light; should be set at .3 or below to avoid a washed out look.

Reflection: color of the reflections or specular reflections.

Roughness: controls size of the reflection. A high value denotes a shiny object which reflects small specular areas. A low value denotes a rough object which has no specular highlights or appears matte.

Transparency: high value, 1, is totally transparent while 0 is solid.

Refraction: controls the refraction index for transparent materials.

Bump Map: controls the the apparent height of the surface of an object to reflect roughness.

1. **Click on "OK" to exit New Standard materials dialog box.**
2. **Click on "Attach..."**
3. **Click on a chair in the perspective viewport to assign fabric to chair.**

You may have to undo the selection if you select the wall or floor by accident.

1. **Render the perspective viewport to view material.**

Modifying Materials

Another way to create materials is just to modify an existing material from the Materials Library. You will duplicate the standard material "aqua glaze," enter a new name, and then modify its attributes.

1. **Click on the Materials**  **icon.**
2. **Click on "Materials Library."**
3. **In the Materials Library, highlight "Aqua Glaze" in the Library List.**
4. **Click on "<-Import" to import to the Materials List.**
5. **Highlight "Aqua Glaze" in the Materials List.**
6. **Click on "OK to exit Materials Library.**
7. **Click on "Duplicate..." (making sure Aqua Glaze is highlighted).**
8. **Type a new Material Name.**
9. **Click on "Preview."**
10. **Change some of the attributes.**
11. **Click on "OK" to exit Modify Standards Materials window.**
12. **Click on "Attach..." and click on an object to assign material.**
13. **Render a scene to see the results.**

Saving Materials in the Library

Once you have created a number of materials, you will want to save them to a new library name so you have a library file containing just the materials you are using on this project. Otherwise, the materials will be lost once you exit AutoCAD.

> **1. Click on the Materials icon.**
> **2. Click on "Materials Library."**
> **3. Click on "Save As..."**
> **4. Type a new name for the library such as "Mine" and put with your file.**
> **5. Click on "Save" under the Materials List.**
> **6. Click on "OK" to exit Materials Library dialogue box.**
> **7. Click on "OK" to exit Materials dialogue box.**

AutoCAD assigns the filetype "mli" to indicate that this file is a materials library. Only the materials under the materials list are saved rather than the original materials library. You can use material library files which have the extension "mli" from other software programs such as 3D Studio Viz or 3D Studio Max also. It is a good idea to have some sort of system for your materials libraries. For instance, you could have one large library of every material you ever created. Or, you could create a library for each project.

In this next section, you will generate another perspective view and put a ceiling on top the room before you learn about mapping.

> **1. Make a box about 1' thick for the ceiling.**
> **2. Use the DVIEW command to create a perspective view so the background is not showing (Fig. 15-4).**

SETUV Command--Mapping

Some materials have bitmap image files. AutoCAD needs to know how to apply these bipmapped images to a surface. It uses four types of mapping to do this: planar, cylindrical, spherical, and solid. For objects shaped like boxes, planar is the most often used mapping. Logically, columns use cylindrical mapping while a round object such as a ball would use spherical mapping. Solid mapping allows you to select the U (width similar to X), V (length similar to Y), and W (depth similar to Z). The best way to learn about them is to try them out. You will use planar mapping to apply a pattern to the floor.

> **1. Click on the Materials icon.**
> **2. Click on "Materials Library."**
> **3. Highlight "Southwest Pattrn" in the Library List.**

4. Click on "<-Import" to import to the Materials List.
5. Highlight "Southwest Patt" in the Materials List.
6. Preview with a cube.
7. Click on "Attach..."

Select objects: **(click on floor)**
Select objects: **(press ENTER)**

8. Click "OK" to close Materials window.
9. Render the scene using Photo Real instead of Render for rendering type and make sure box next to Apply Materials is checked.

The mapping by default is planar and the southwest pattern is mapped 1:1 over the whole floor (Fig. 15-6). Basically, AutoCAD maps the image pixel by pixel. If you had not used photorealistic, you would not see the patterned floor.

Fig. 15-6 Rendered image with patterned floor.

In the next section, you will apply a tile map to the floor to repeat the pattern.

1. Click on the Mapping  **icon.**
2. Select floor.
3. Click on "Adjust Coordinates..."

You will notice that "WCS XY Plane" is checked for the Parallel Plane meaning that the World Coordinate system and the XY plane will be used for the mapping. That's fine for the floor which is parallel to the WCS. You would use the XZ plane if you were mapping to the back wall while you would use the YZ plane to map the side walls.

In the Center Position area of the Adjust Planar Coordinates dialogue box you see the mapping icon which is aqua and has a little handle. It should be centered over the floor but if not, you can adjust with the XY sliders. This indicates that the center of the bitmap image will be in the center of the floor.

In the Offsets and Rotations area, you can also offset the image by entering coordinates which is similar to using sliders to center the position. You can also enter an angle to rotate the image.

4. Try adjusting the center position or rotation.
5. Preview to see the results.
6. Adjust map to center of floor.
7. Click on the "Adjust Bitmap..."

The Adjust Object Bitmap Placement dialogue box appears. In this box you can adjust the scale and tile the image.

8. Check Tile.
9. Enter 2 for both the U and V scale.
10. Click on "Preview."

You should see four repetitions of the pattern. U and V are the coordinates used to describe the horizontal and vertical directions of the bipmap to distinguish from the WCS. A map has its own U and V coordinates independent of the WCS or UCS which can be rotated or scaled.

11. Enter 4 for both U and V scale.
12. Click on "Preview."
13. Click on "OK" three times to go back to AutoCAD.
14. Render the scene.

You should see the pattern repeated about 16 times (Fig. 15-7). Actually, the pattern is repeated more in the V direction since the floor is not square.

Fig. 15-7 Image using 4 x 4 scale for the mapping.

Using Fog

Fog is most useful in rendering outdoor scenes and you can get interesting results with interiors also. You can access the fog dialogue box directly from the Render dialogue menu.

1. Click on the Render icon.
2. Click on the "Fog/Depth" button and adust the fog parameters.
3. Render the scene.

Saving Your Rendering

Each time you rendered a scene you displayed the image on the screen, but it was not saved. AutoCAD saves the image to a bimap or "bmp" file format. You can change this format easily to other file formats (TIFF, TGA) that are used more often in other programs.

1. Click on the Render icon.
2. Select "File" for Destination.
3. Click on "More Options...."
4. Select "TIFF" for File type.
5. Select "2048 x 1366" for resolution.
6. Click on "OK" to exit to Render dialogue box.
7. Click on "Render" to render to a file.

You rendered to a higher resolution which will take more time. This file can be loaded into a program such as Adobe Photoshop and printed. TIF filetype works very well for printing.

Printing Rendered Images from AutoCAD

You can also print an image that is displayed in a viewport directly from AutoCAD. Of course, the resolution will be lower because it is dependent on the screen resolution.

1. Click on the Render icon.
2. Select "Render Window" for Destination.
3. Click on "Render" to display in the render window.
4. Click on the printer icon to print.

If you did not get a print, then the printer was probably not configured with Windows. Generally, you will have better results saving to a file and printing with another program.

Notes on Rendering

This tutorial has introduced you to AutoCAD rendering features. You can also export the geometry to rendering and animation programs such as 3D Studio Viz or Max by using 3DSOUT command. Additionally, you can import the geometry from a 3D Studio file by using 3DSIN command. These commands can be most easily entered by using the File menu and clicking on "Export...: Note that 3D Studio uses named objects so it is best to organize your drawing by layers putting an object on each layer with each name. Then, when 3D Studio Viz or Max imports the drawing, you will have the option of using the layers for object names.

TUTORIAL 16

SOLID MODELING

Commands Learned:	BOX	SUBTRACT	CHAMFER	REVOLVE
	CYLINDER	UNION	EXTRUDE	SOLIDEDIT

In Tutorials 12 through 14, you learned how to do surface modeling. In this tutorial you will learn how to do solid modeling by making a room in an art museum.

You can think of surface modeling as a wireframe with a mesh draped over it—hence surface modeling. An object created with solid modeling is solid with properties of volume and mass. A designer is mainly concerned with visual effects of solid modeling rather than these properties. However, solid modeling offers advantages over surface modeling in editing. In contrast to surface modeling, you can modify primitive shapes of the box, cylinder, cone, etc. into more complex shapes through the use of boolean geometry or more advanced solid editing. Boolean operations allow you to join solids, subtract solids from another, or create a new object from the intersection of two solid objects.

1. Menu Bar: File > New...
2. Click on "Use a Wizard" and "Quick Setup."
3. Click on "Architectural" for units.
4. Enter 84' for width and 96' for length of area.
5. Click "Finish"button.
6. Zoom all.

You will view the Solids and Solids Editing toolbar to use for this tutorial.

1. Close the dimensioning toolbar if it is still displayed.
2. Right click on any icon and check "Solids" to view Solids toolbar.
3. Right click on any icon and check "Solids Editing" to view Solids Editing toolbar.
4. Place these toolbars on the right side of the drawing area.

In surface modeling, you created several UCS, User Coordinate Systems, or drawing planes in addition to using the WCS, World Coordinate System. The WCS and UCS are equally important in solid modeling. AutoCAD draws objects parallel to the drawing plane or UCS. Drawing objects on the floor requires use of WCS and doing your work in plan view. Drawing objects on the walls requires you to draw in the elevation view with the UCS parallel to the wall.

BOX Command

The BOX command draws solid boxes which can be edited in various ways. You will use it to create the floor and two walls of the room.

1. Set SNAP to 6."

Command: BOX
Specify corner of box or [CEnter] <0,0,0>: **(Click in lower left corner of screen)**
Specify corner or [Cube/Length]: **L**
Specify length: **30'**
Specify width: **24'**
Specify height: **-1'**

1. Zoom so you see floor and just a little area around it.

You specified -1 feet so you will not have to set the elevation to 1 feet this time. You will first draw the west wall.

Command: BOX
Specify corner of box or [CEnter] <0,0,0>:**(Click lower left corner of the floor)**
Specify corner or [Cube/Length]: **L**
Specify length: **6"**
Specify width: **23'6"**
Specify height: **15'**

Next you will draw the west wall:

Command: BOX
Specify corner of box or [CEnter] <0,0,0>:**(Click on upper left corner of west wall or wall you just drew)**

Specify corner or [Cube/Length]: **L**
Specify length: **30'**
Specify width: **6"**
Specify height: **15'**

1. Menu Bar: View > Viewports > 4 Viewports.
2. Click the lower left viewport to make it active.
3. Menu Bar: View > 3D Views > Left.
4. Click the lower right viewport to make it active.
5. Menu Bar: View > 3D Views > Front.

6. Click upper right viewport to make it active.
7. Menu Bar: View > 3D Views > SE Isometric.

The plan view is in the upper left viewport, left elevation in the lower left viewport, front elevation in the lower right viewport, and the SE Isometric in the upper right viewport (Fig. 16-1).

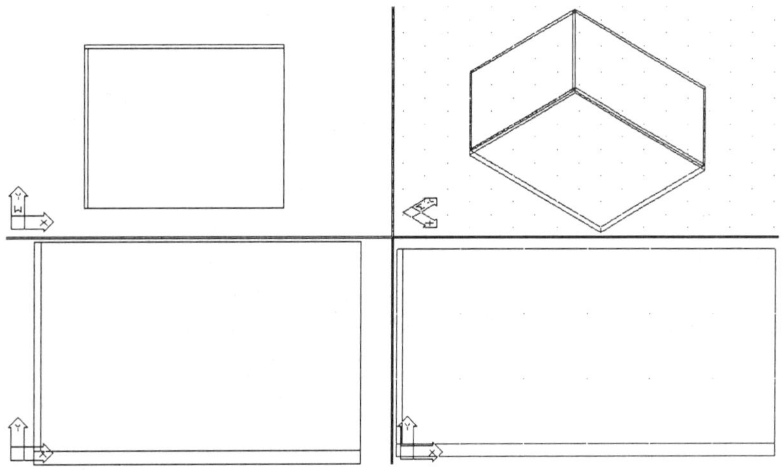

Fig. 16-1 Viewport configuration showing plan, elevation and isometric views.

You will save this viewport configuration:

1. Menu Bar: View > Viewports > New Viewports.
2. Make sure "Active Model Configuration" is highlighted.
3. Type "4views" in box next to "New name:"
4. Click "OK" to close Viewports window.

CYLINDER Command

The CYLINDER command is very useful combined with boolean operations. In the next task you will make a box and cylinder. These will be used to subtract from the west wall to make an arched door opening so you will move that wall out to facilitate the operation.

1.Make sure ORTHO is on and SNAP is on and set at 6."
2. In the plan view or upper left viewport, move the west wall to the west about 7-10 feet.

Command: BOX
Specify corner of box or [CEnter] <0,0,0>: **(click anywhere on floor in plan view)**
Specify corner or [Cube/Length]: **L**
Specify length: **2'**
Specify width: **6'**
Specify height: **7'**

Your box should appear similar to the illustration (Fig. 16-2).

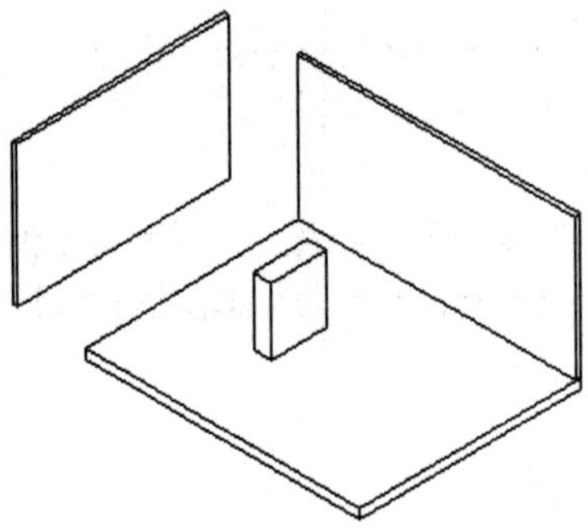

Fig. 16-2 Box to be used for boolean.

2. Click in the upper right viewport to make it active.
3. Turn off grid by press F7 key.

Command:**UCS**
Current ucs name: *WORLD*
Enter an option [New/Move/orthoGraphic/Prev/Restore/Save/Del/Apply/?/World]
<World>: **X**
Specify rotation angle about X axis <90>: **90**
Command: **(press ENTER)**
UCS
Current ucs name: *NO NAME*
Enter an option [New/Move/orthoGraphic/Prev/Restore/Save/Del/Apply/?/World]
<World>: **Y**
Specify rotation angle about Y axis <90>: **90**
Command: **(press ENTER)**
UCS
Current ucs name: *NO NAME*
Enter an option [New/Move/orthoGraphic/Prev/Restore/Save/Del/Apply/?/World]
<World>: **O**
Specify new origin point <0,0,0>: **APPINT (or click on** 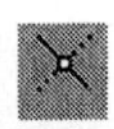 **icon)**
of **(click on lower left front corner of box in upper right viewport)**

The UCS icon should be on the lower left corner and aligned with the front of the box. If not, undo the 3 commands and try again. Next, you will draw a cylinder on top of the box.

Command:
CYLINDER
Current wire frame density: ISOLINES=4
Specify center point for base of cylinder or [Elliptical] <0,0,0>: **mid (or** 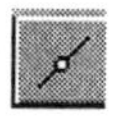 **icon)** of **(click on top line of box)**
Specify radius for base of cylinder or [Diameter]: **3'**
Specify height of cylinder or [Center of other end]: **-2'**

1. If the cylinder is not directly on top of the box as illustrated, move the cylinder in plan view to be on top of the box (Fig. 16-3).

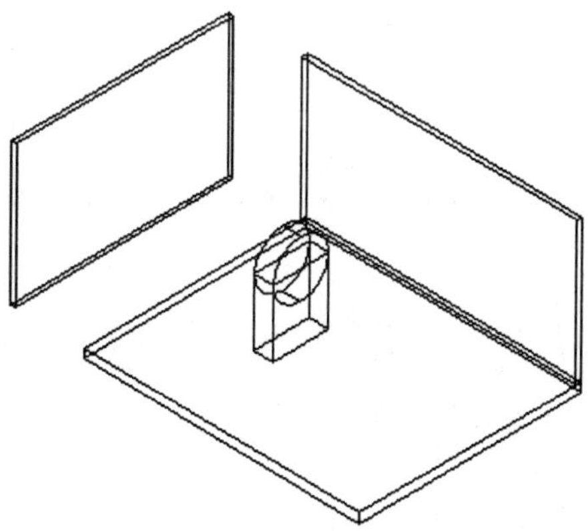

Fig. 16-3 Box with cylinder on top.

Boolean Operations

Boolean geometry allows you to join one solid shape to another, subtract one from the other, and use the intersection between the two objects to create a new object. In the following illustration (Fig. 16-4), you can see the results of the three boolean operations on two overlapping circles.

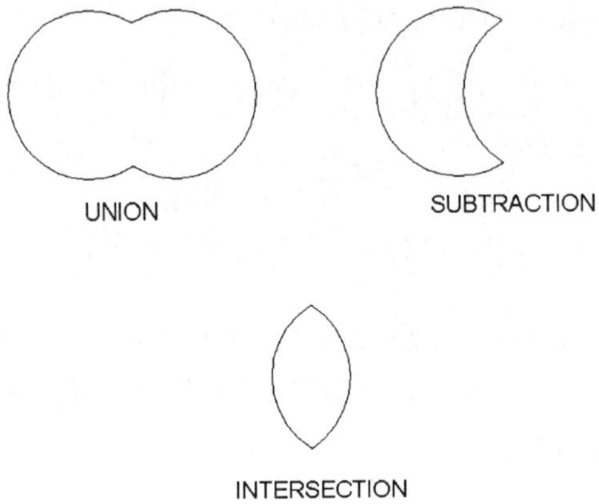

Fig. 16-4 The three boolean operations.

Subtraction is useful for making holes in the walls for windows or doors. It is best to make sure that an object does not overlap an object that you do not want to boolean. You have moved the west wall out so it does not interfere with the floor.

1. Move both the box and cylinder to overlap the west wall.

In plan view, your doorhole should look as illustrated (Fig. 16-5). In the elevation, check to make sure that the doorhole goes all the way to the bottom so when you subtract you will not get a little piece at the bottom of the wall (Fig. 16-6).

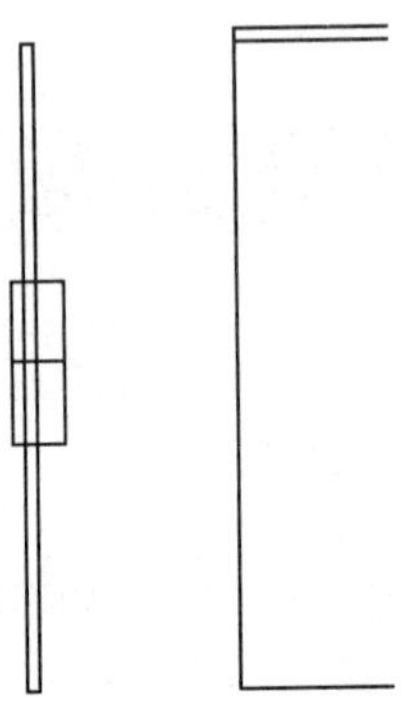

Fig. 16-5 Plan view.

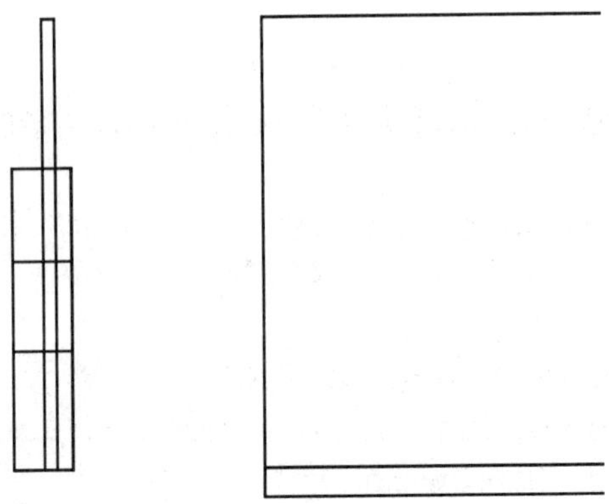

Fig. 16-6 Elevation view.

Now, you are ready to subtract the two objects from the wall.

 **icon or enter:**

Command: SUBTRACT
Select solids and regions to subtract from.
Select objects: **(Click on west wall)**
Select objects: **(Press ENTER or right click)**
Select solids and regions to subtract...
Select objects: **(Click on both box and cylinder)**
Select objects: **(Press ENTER or right click)**

Your wall should appear as illustrated (Fig. 16-7).

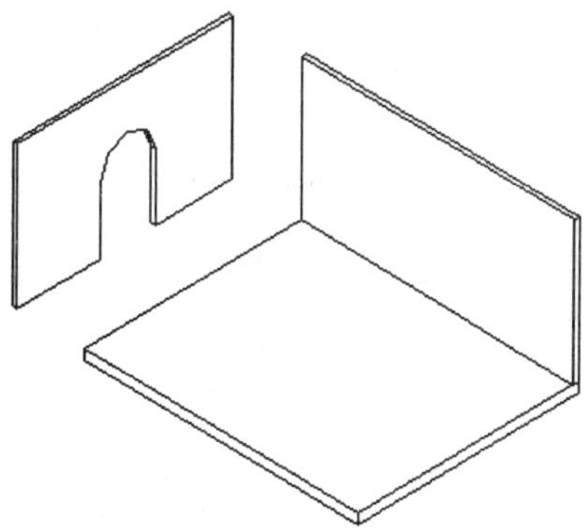

Fig 16-7 Door opening after subtraction boolean.

1. Click on the upper right viewport or SE Isometric view.

The UCS is still rotated. You will return the UCS to WCS but first save this UCS in case you might need it again.

Command: **UCS**

Current ucs name: *NO NAME*
Enter an option [New/Move/orthoGraphic/Prev/Restore/Save/Del/Apply/?/World]
<World>: **S**
Enter name to save current UCS or [?]: **XYAXIS**
Command: **(press ENTER)**
UCS
Current ucs name: XYAXIS
Enter an option [New/Move/orthoGraphic/Prev/Restore/Save/Del/Apply/?/World]
<World>:**(press ENTER)**

1. Make sure ORTHO is on and SNAP is on and at 6."
2. Move the wall back to the room.

Next, you will make a pedestal for artwork by joining a rectangular solid with two cylinders. First you need to make and place the objects as shown (Fig. 16-8).

1. Click in the upper left or plan viewport to make it active.
2. Use BOX command to make box 8' feet in length, 2' in width, and 3' high.
3. Make a solid cylinder and place on the end of the box (use mid to make sure it will align with end) using 1' for radius and same 3' for height.
4. Make copy of the cylinder and move to the other end (Fig. 16-8, 16-9).
5. Click on the upper right viewport to make it current.
6. Use the HIDE command or click on the  icon on the Render toolbar.

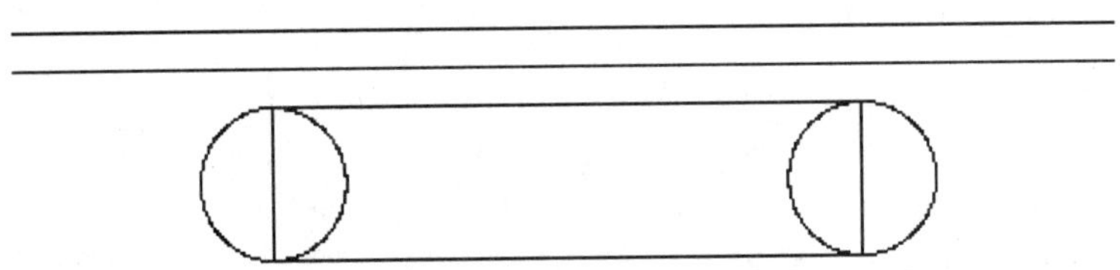

Fig. 16-8 Box with two cylinders.

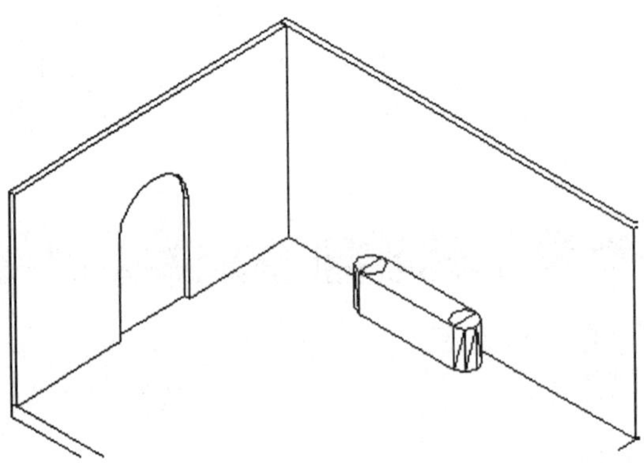

Fig. 16-9 Pedestal before boolean operation.

You can see there is overlapping geometry of the cylinders and the box which may cause errors in rendering. Generally, overlapping geometry is a bad practice while clean geometry is considered to be the hallmark of a good computer modeler. Next, you will join the objects to make one solid object by using a union operation.

1. Turn SNAP off.

Command: UNION
Select objects: **(select cylinder)**
Select objects: **(select cylinder)**
Select objects: **(select box)**
Select objects: **(Press ENTER or right click)**

1. Click on the upper right viewport to make it current.
2. Use HIDE command.

You can see that the pedestal appears correct now (Fig. 16-10).

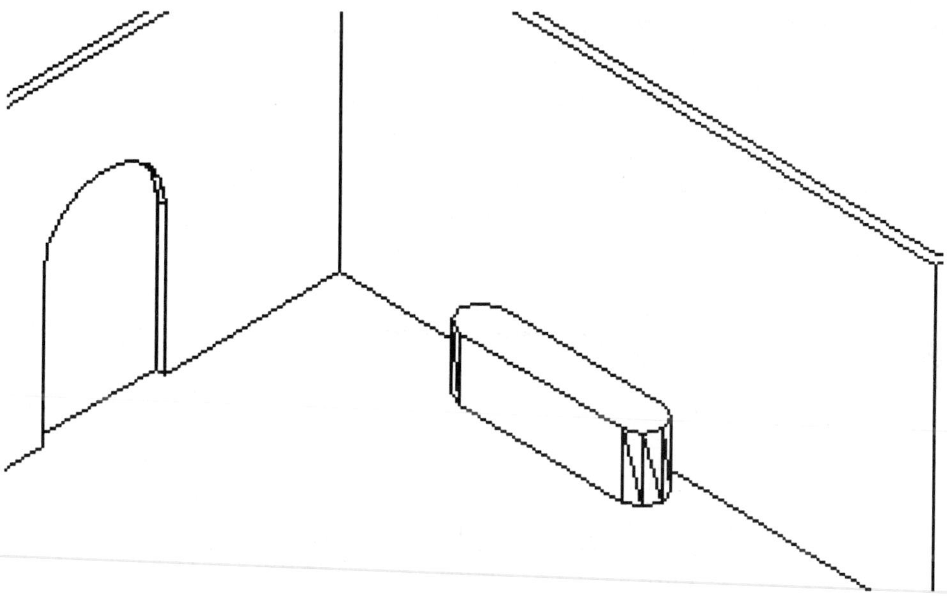

Fig. 16-10 Pedestal after boolean operation.

CHAMFER Command

Some of the edit commands such as FILLET and CHAMFER can be used on solid objects with interesting results. The best way to learn about them is by trial and error.

1. **Create a couple of boxes anywhere in your room.**
2. **Use the FILLET command to curve surfaces of the box.**

You will use the CHAMFER command to cut off one surface of the second box by entering:

Command: CHAMFER
(TRIM mode) Current chamfer Dist1 = 0'-0 1/2", Dist2 = 0'-0 1/2"
Select first line or [Polyline/Distance/Angle/Trim/Method]: **(select top of box in the upper right viewport)**

Base surface selection...
Enter surface selection option [Next/OK (current)] <OK>: **(press ENTER)**
Specify base surface chamfer distance <0'-0 1/2">: **6**
Specify other surface chamfer distance <0'-0 1/2">: **2**
Select an edge or [Loop]: **(select the same top edge)**
Select an edge or [Loop]: **(press ENTER)**

Your boxes may look as illustrated (Fig. 16-11).

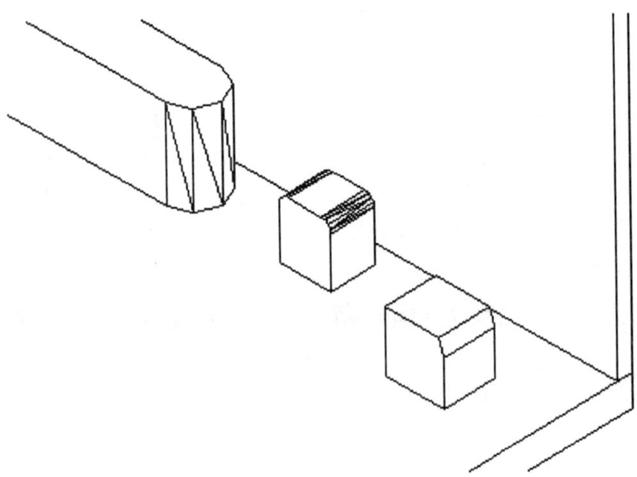

Fig. 16-11 Boxes after use of FILLET and CHAMFER.

EXTRUDE Command

The EXTRUDE command allows you to create solids by extruding a two-dimensional object. The objects must be closed and can include circles, polygons, rectangles, ellipses, polylines, or splines. You have the option of either extruding along a path or adding height and a taper angle. You will do both of these in the next sequence of tasks.

1. Draw an ellipse.
2. Make a copy of the ellipse and move a few feet away.
3. Click on the Extrude icon or enter:

Command: EXTRUDE
Select objects: **(select one ellipse)**
Select objects: **(press ENTER or right click)**
Path/<Height of Extrusion>: **6'**
Extrusion taper angle <0>: **(press ENTER or right click)**

For the next shape, you need to create a path which can be done by drawing a polyline and editing it to make it curve. You will need to rotate the UCS to draw the polyline in the front elevation view.

1. Click on the icon to rotate the UCS 90 degrees around the X axis.
2. Turn SNAP on.

You should see the X-Y arrows on the UCS icon indicating it is parallel to the front elevation in that viewport.

3. Use the PLINE command and start in the plan viewport using osnap to snap to the center of the ellipse.
4. Click in the front elevation viewport and continue with several points for the polyline.
5. Use the PEDIT command with the Fit or Spline option to make a smooth curve (Fig. 16-12.)

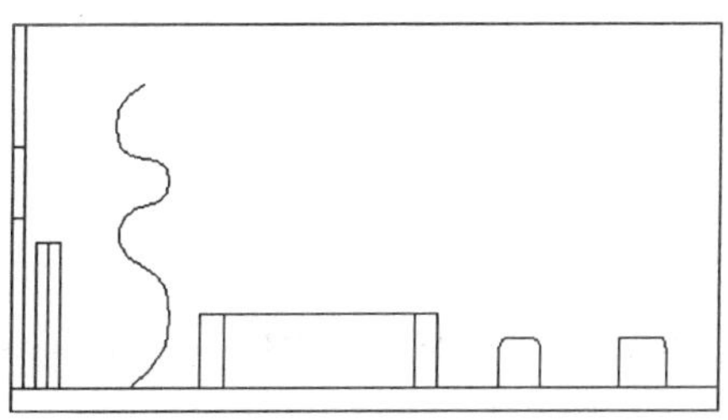

Fig. 16-12 Curved polyline to be used for path.

Now you will extrude the second ellipse along this path.

1. **Return the UCS to the WCS.**
2. **Click on the Extrude icon or enter:**

Command: EXTRUDE
Select objects: **(select the second ellipse)**
Select objects: **(press ENTER or right click)**
Path/<Height of Extrusion>: **P**
Select path: **(select curved polyline)**

1. **Use the HIDE command or click on the icon on the Render toolbar.**

AutoCAD may have rotated your ellipse to do the extrusion. Your ellipses may look like the illustration (Fig. 16-13).

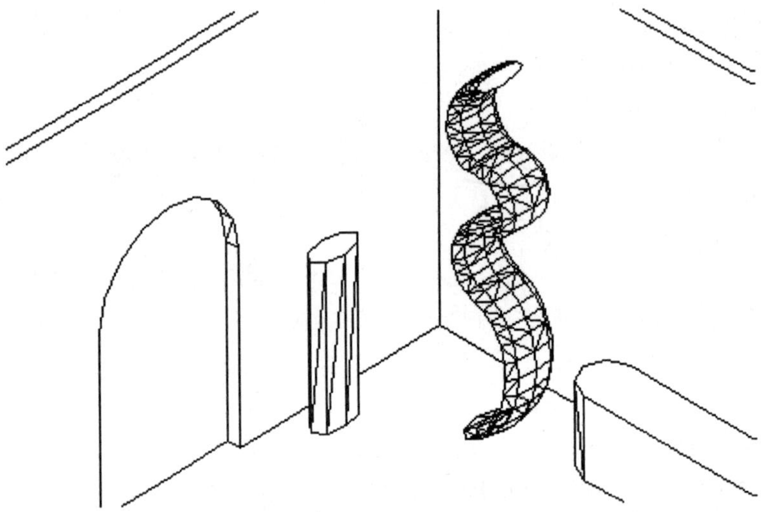

Fig. 16-13 Extruded ellipses.

REVOLVE Command

The REVOLVE command is very similar to the REVSURF command you used earlier. The difference is that you can make solid revolved objects with REVOLVE. You will use it to make a large columun in the center of the room.

1. Click on the icon to rotate the UCS 90 degrees around the X axis.

2. Use the PLINE command to draw a closed polyline in the front viewport a little above the floor (Fig. 16-14). Note you can use arcs and lines and join and convert them to polylines with the PEDIT command.

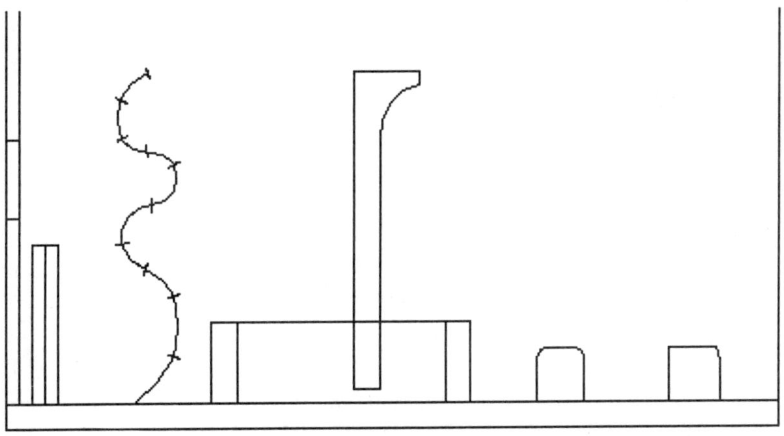

Fig. 16-14 Closed polyline.

3. Click on the icon or enter:

Command: REVOLVE
Select objects: **(click on polyline)**
Select objects: **(press ENTER or right click)**
Axis of revolution - Object/X/Y/<Start point of axis>: **(click inside bottom corner of polyline)**
<End point of axis>: **(click inside top corner of polyline)**
Angle of revolution <360>: **(press ENTER or right click)**

1. Move the column back down to the floor level.
2. Change the UCS back to WCS.
3. Move the column back into the room in the floor plan viewport.

The column should appear similar to what is illustrated (Fig. 16-15).

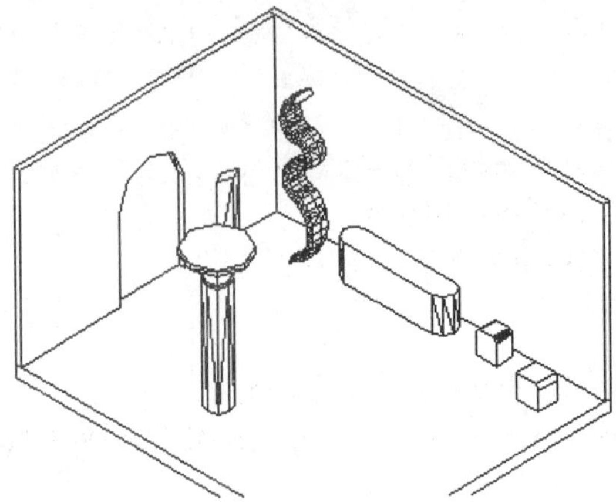

Fig. 16-15 Revolved column.

4. Continue making solid objects to finish the room.

In the illustration, the pyramid was made with surface modeling (Fig. 16-16). The other objects were copied. This was rendered with two spotlights and a point light.

Fig. 16-16 Rendered room.

Advanced Work with Solids and Editing

For the most part, you will use the primitive solids and boolean operations to create composite solids for interiors or architectural models. However, there may be times you want to do more advanced solid editing, particularly for furniture design or complex architectural models. AutoCAD 2000 provides some powerful tools for this advanced work. These are best learned by practicing on basic solids. You will create a new drawing and make several solid boxes to practice these edit tools.

1. **Menu Bar: File > New...**
2. **Click on "Use a Wizard" and "Quick Setup."**
3. **Click on "Architectural" for units.**
4. **Enter 50' for width and 30' for length of area.**
5. **Click "Finish" button.**
6. **Zoom all.**
7. **Make a solid box about 5' by 3' x 4' high.**
8. **Use COPY command to make two copies of the box spaced 8' apart.**

Now you will create a viewport layout to see your work:

1. **Menu Bar: View > Viewports > 3 Viewports.**
2. **Click the lower left viewport to make it active.**
3. **Menu Bar: View > 3D Views > Front.**
4. **Click the right viewport to make it active.**
5. **Menu Bar: View > 3D Views > SE Isometric.**

SOLIDEDIT Command

The SOLIDEDIT command edits composite solids or the objects you created by boolean operations. In previous AutoCAD releases, composite solids were not able to be edited in this way. You could only do additional boolean operations.

SOLIDEDIT has many options that allow you to precisely refine any objects. Each option has its own icon on the Solids Editing toolbar. After entering the command, you are first prompted to select which part of the object is to be edited. Each part then has several editing options. The parts or objects include:

Face: Plane or planar surface which is rectangular, trianglular, cylindrical, or curved.

Edge: Edge created by intersection of two faces and has two-dimensional appearance such as a line, arc. or circle.

Body: A 3D solid object.

SOLIDEDIT Command - Face Taper Option

The Taper option is used to change the angle of a face or plane.

Command: **SOLIDEDIT**
Solids editing automatic checking: SOLIDCHECK=1
Enter a solids editing option [Face/Edge/Body/Undo/eXit] <eXit>: **F**
Enter a face editing option
[Extrude/Move/Rotate/Offset/Taper/Delete/Copy/coLor/Undo/eXit] <eXit>: **T**
Select faces or [Undo/Remove]: **(click in front view or click on top edge of box).**

2 faces found.
Select faces or [Undo/Remove/ALL]: **(press ENTER)**
Specify the base point: **(click lower left corner of face)**
Specify another point along the axis of tapering: **(click upper left corner of face)**

Specify the taper angle: **30**
Solid validation started.
Solid validation completed.
Enter a face editing option
[Extrude/Move/Rotate/Offset/Taper/Delete/Copy/coLor/Undo/eXit] <eXit>:**(press ENTER)**

Solids editing automatic checking: SOLIDCHECK=1
Enter a solids editing option [Face/Edge/Body/Undo/eXit] <eXit>: **(press ENTER)**

The result may be similar to the illustration (Fig. 16-17). If you had clicked on the  icon instead of typing the command, the first two options would already been entered and you would have been prompted to select faces. You specified a postive angle for the taper which resulted in tapering the face inward or forward. If you had specified a negative angle, the face would be outward or away from the object.

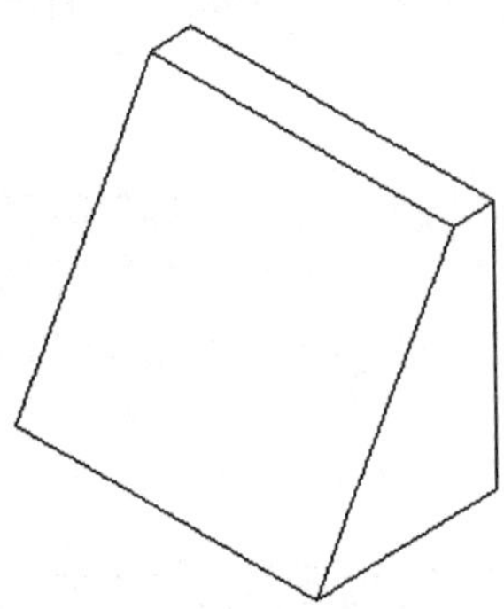

Fig. 16-17 Face of box tapered.

SOLIDEDIT Command - Face Move Option

The Move option will move one part of the 3D solid without affecting the other parts which is similar to STRETCH command in 2D work.

1. In the middle of the next box, draw a cylinder, 1' in radius, and 5' high so it is taller than box.
2. Use SUBTRACT command and substract the cylinder from the box to make a hole.
3. Make a copy of the box with the hole in it.

You will move the hole to one end of the box. This time, you will use the icon to cut down on the typing:

Select faces or [Undo/Remove]: **(select hole in front view)**
Select faces or [Undo/Remove/ALL]: **(press ENTER)**
Specify a base point or displacement: **(click on hole in front view)**
Specify a second point of displacement: **(move to right in front view)**
Solid validation started.
Solid validation completed.
Enter a face editing option
[Extrude/Move/Rotate/Offset/Taper/Delete/Copy/coLor/Undo/eXit] <eXit>: **(press ENTER)**
Solids editing automatic checking: SOLIDCHECK=1
Enter a solids editing option [Face/Edge/Body/Undo/eXit] <eXit>:**(press ENTER)**
Command: **HIDE**

The result should be similar to the view on the right (Fig. 16-18).

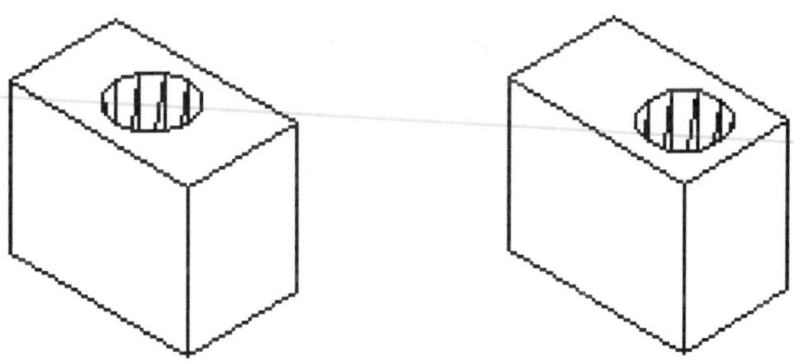

Fig. 16-18 View before and after move option.

The Face Rotate option is similar. If the hole was a different shape, square or ellipse, you could have rotated it inside the box.

SOLIDEDIT Command - Face Offset Option

Similar to the OFFSET command, the offset option of SOLIDEDIT creates a 3D offset but does not make a copy of the object. You will try it on the box with the original hole location to make the hole smaller.

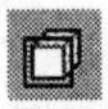

Select faces or [Undo/Remove]: **(select hole in front view)**
Select faces or [Undo/Remove/ALL]: **(press ENTER)**
Specify the offset distance: **3**
Solid validation started.
Solid validation completed.
Enter a face editing option
[Extrude/Move/Rotate/Offset/Taper/Delete/Copy/coLor/Undo/eXit] <eXit>: **(press ENTER)**
Solids editing automatic checking: SOLIDCHECK=1
Enter a solids editing option [Face/Edge/Body/Undo/eXit] <eXit>:**(press ENTER)**
Command: **HIDE**

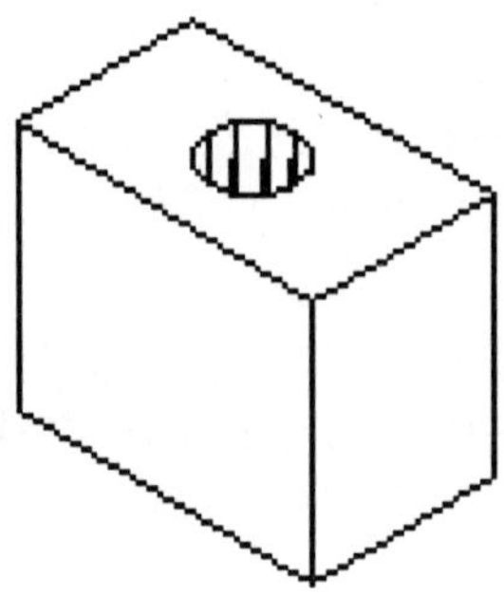

Fig. 16-19 Offset creates smaller hole.

The 3 inch offset creates a smaller hole 18" in diameter in contrast to the original 24" diameter hole. A negative offset value would enlarge the hole.

SOLIDEDIT Command - Face Extrude Option

The Extrude option allows you to extrude a face of the composite solid to a specified height or along a path. In this option, you must select a single face. By clicking anywhere on the box, you select two faces so you will use the Remove option in order to remove the face so that you have only the top face selected .

1. Use COPY to make a copy of the box with the moved hole (Fig. 6-20).

(on Solids Editing toolbar)
Select faces or [Undo/Remove]: **(Select top face in SE isometric view)**
Select faces or [Undo/Remove/ALL]: **R**
Remove faces or [Undo/Add/ALL]: **(select vertical face so that top face is only one selected)**
Remove faces or [Undo/Add/ALL]: **(press ENTER)**
Specify height of extrusion or [Path]: **7'**
Specify angle of taper for extrusion <0>: **(press ENTER)**
Enter a face editing option
[Extrude/Move/Rotate/Offset/Taper/Delete/Copy/coLor/Undo/eXit] <eXit>: **(press ENTER)**
Solids editing automatic checking: SOLIDCHECK=1
Enter a solids editing option [Face/Edge/Body/Undo/eXit] <eXit>:**(press ENTER)**
Command: **HIDE**

The box should be 7' high or 3' higher than the original box (Fig. 16-20). If the hole had been square, you could select one of the vertical faces for extrusion which would result in opening up one side of the object.

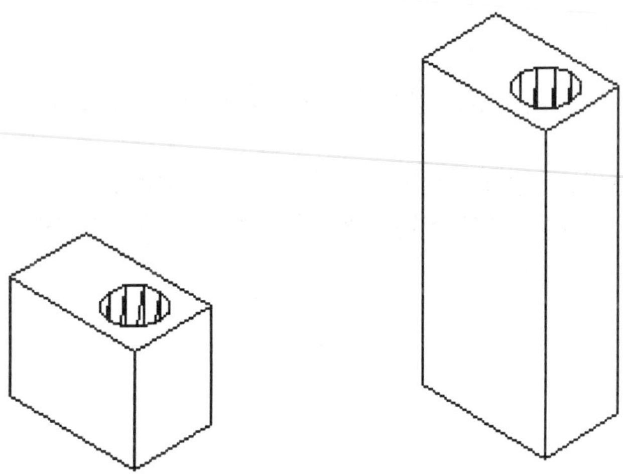

Fig. 16-20 Composite solid extruded to make taller object.

SOLIDEDIT Command - Face Copy Option

This option creates a two-dimensional copy of a face of the object. Once again, you probably need to use the Remove option to only select the top face. For this task, use the box that you just made a copy of or the box with the moved hole.

Select faces or [Undo/Remove]: **(Select top face of box with hole in it)**
Select faces or [Undo/Remove/ALL]: **R**
Remove faces or [Undo/Add/ALL]: **(select vertical face so that top face is only one selected)**
Remove faces or [Undo/Add/ALL]: **(press ENTER)**
Specify a base point or displacement: **0,0,0**
Specify a second point of displacement: **0,4',0**
Enter a face editing option
[Extrude/Move/Rotate/Offset/Taper/Delete/Copy/coLor/Undo/eXit] <eXit>: **(press ENTER)**
Solids editing automatic checking: SOLIDCHECK=1
Enter a solids editing option [Face/Edge/Body/Undo/eXit] <eXit>:**(press ENTER)**

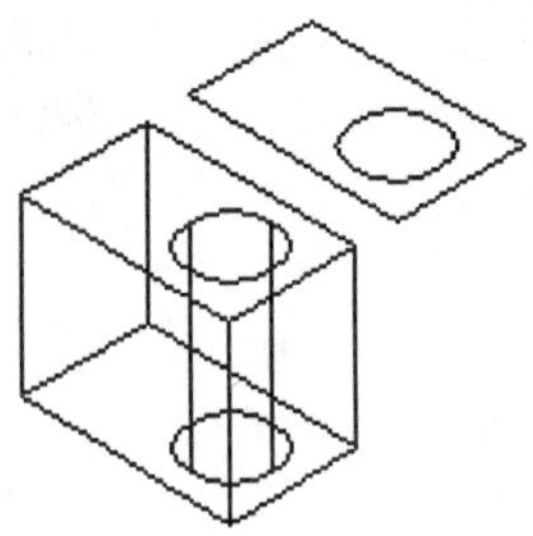

Fig. 16-21 Copied face.

The Extrude Face option will not work on this face so you will use the EXTRUDE command.

1. Use the EXTRUDE command with -2' as extrusion height (Fig. 16-22).

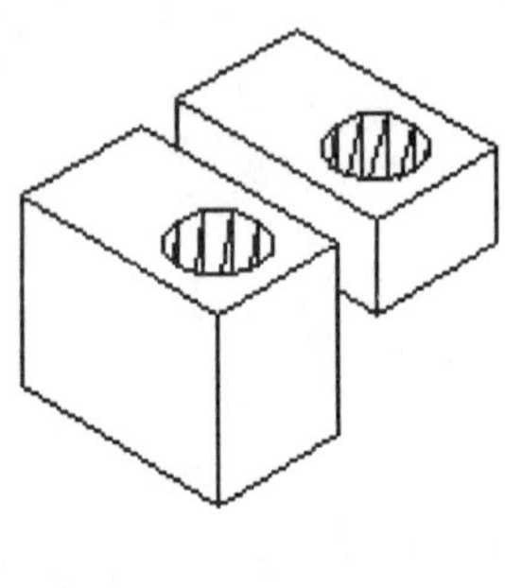

Fig. 16-22 Extruded face.

SOLIDEDIT Command - Face Delete Option

The delete option allows you to delete a face on the solid object. This can have unexpected results on complex objects or will not be able to be done on simple objects. You will use it to remove the hole in the box when you used the offset option.

1. **Use the COPY command to make a copy of the box with the 18" diameter hole.**

Select faces or [Undo/Remove]: **(select hole in front view)**
Select faces or [Undo/Remove/ALL]: **(press ENTER)**
Enter a face editing option
[Extrude/Move/Rotate/Offset/Taper/Delete/Copy/coLor/Undo/eXit] <eXit>: **(press ENTER)**

Solids editing automatic checking: SOLIDCHECK=1
Enter a solids editing option [Face/Edge/Body/Undo/eXit] <eXit>:**(press ENTER)**

The hole is removed leaving the original box (Fig. 16-23).

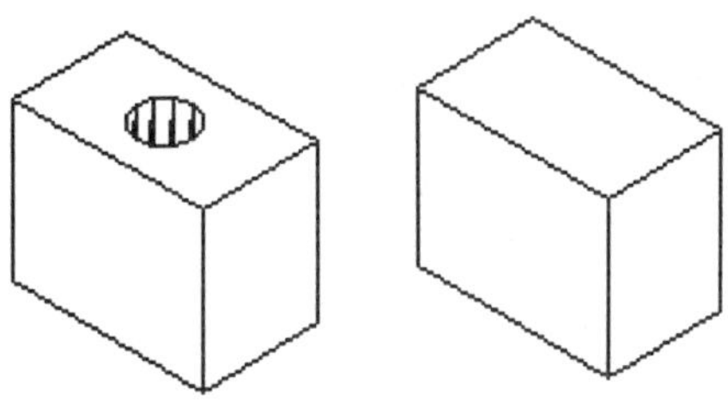

Fig. 16-23 The hole is eliminated with the face delete option.

SOLIDEDIT Command - Edge Copy Option

There are only two edge options, copy and color. The copy option will always result in a two-dimensional shape. You will use this option to make a copy of the hole which will make a circle:

Select edges or [Undo/Remove]: **(select 18” dimeter hole on last box)**
Select edges or [Undo/Remove]: **(press ENTER)**
Specify a base point or displacement: **0,0,0**
Specify a second point of displacement: **0,0,7'**
Enter an edge editing option [Copy/coLor/Undo/eXit] <eXit>: **(press ENTER)**
Solids editing automatic checking: SOLIDCHECK=1
Enter a solids editing option [Face/Edge/Body/Undo/eXit] <eXit>: **(press ENTER)**

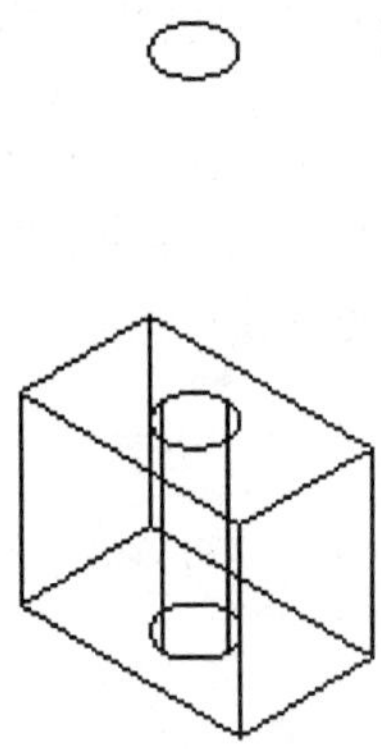

Fig. 16-24 Circle created by copying edge of hole.

Of course, that is no difference between this shape and another shape created with the circle command. Had the hole been a more complex shape, you could have created a copy of it which might be difficult to recreate otherwise.

SOLIDEDIT Command - Imprint Option

So far, you have used faces or edges for editing. Now you will use body options which allows you to edit the three-dimensional solid. The Imprint option allows you to attach a two-dimensional shape onto a face of a three-dimensional solid. You will first make a six-sided polygon on top of the original box.

1. Use the ELEV command to set the elevation to 4 feet.
2. Use POLYGON command to make a 6-sided polygon and place on box making sure it cantilevers over the edge (Fig. 16-25).

Select a 3D solid: **(select box)**
Select an object to imprint: **(select polygon)**
Delete the source object <N>: **Y**
Select an object to imprint: **(press ENTER)**
Enter a body editing option
[Imprint/seParate solids/Shell/cLean/Check/Undo/eXit] <eXit>: **(press ENTER)**
Solids editing automatic checking: SOLIDCHECK=1
Enter a solids editing option [Face/Edge/Body/Undo/eXit] <eXit>: **(press ENTER)**
Command: **HIDE**

You can see that the polygon has been cropped to fit the solid box (Fig. 16-25).

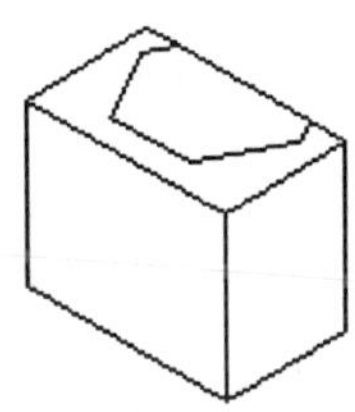

Fig. 16-25 Imprinted polygon.

1. Use SOLIDEDIT with extrude option to make a solid feature of imprint (Fig. 16-26).

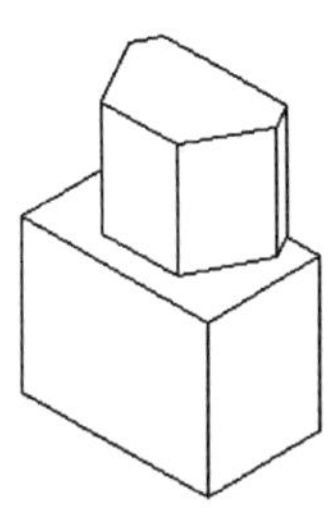

Fig. 16-26 Extruded shape of imprint.

SOLIDEDIT Command - Shell Option

The Shell option will make a wall or shell out of an object. You will use it to make a shell of of the original box.

1. Make a copy of one of the solid or original boxes (one without hole).

Select a 3D solid: **(select a solid box)**
Remove faces or [Undo/Add/ALL]: **(remove the front face from selection set by clicking top and bottom edges of front face as in Fig. 16-26)**
Remove faces or [Undo/Add/ALL]: **(press ENTER)**
Enter the shell offset distance: **3"**
Solid validation started.
Solid validation completed.
Enter a body editing option
[Imprint/seParate solids/Shell/cLean/Check/Undo/eXit] <eXit>: **(press ENTER)**
Solids editing automatic checking: SOLIDCHECK=1
Enter a solids editing option [Face/Edge/Body/Undo/eXit] <eXit>:**(press ENTER)**
Command: **HIDE**

These seult should look like illustration (Fig. 16-27).

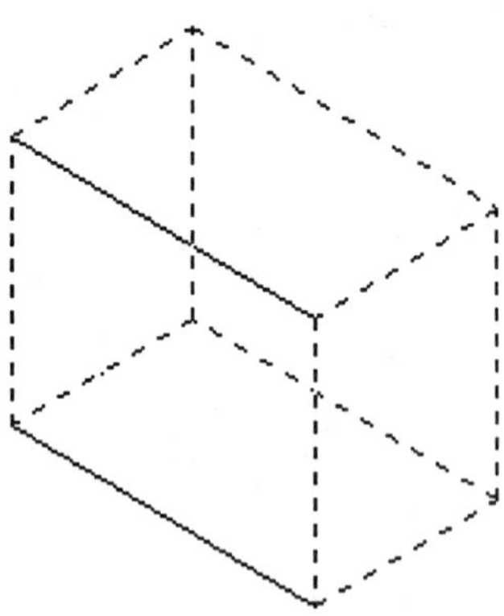

Fig. 16-26 Select faces except front.

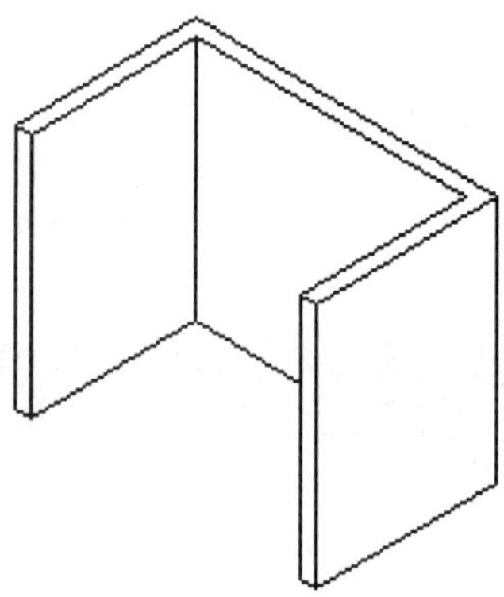

Fig. 16-27 Final shell of object.

Other SOLIDEDIT Options

Separate. The separate option indicated by the icon is useful when you need to separate two unattached objects that because of boolean operations became one object. This will not work on attached objects however, that is those touching each other.

Check. This option indicated by the option will check the geometry to see if it is valid.

Clean. If an error is found such as intersecting geometry, the Clean option indicated by the icon may be able to fix the error.

Notes on Modeling and Visualization

AutoCAD 2000 has many sophisticated modeling features that you may find it a little bewildering. However, you will find there are many ways to model. Some designers like to use boolean operations because they may find them more intuitive. Others may find the SOLIDEDIT command more efficient and easier to conceptualize. No matter which commands you use, it is critically important that you be able to visualize the modeling operations. Often it will take many steps to create an object. To help you figure out what commands to use and in what sequence, you can sketch the various steps by hand on paper. In fact, quick sketching skills are just as important as they have ever been for designers. Likewise, it is important to be able to visualize your concept in your head before you can put it down on paper.

Where Do You Go From Here?

This book has provided you with an introduction to AutoCAD. Now it is up to you to apply your skills to projects where you will increase your productivity and eventually acquire proficiency. Many tools are available that can help you with this endeavor. These include more AutoCAD references, symbol libraries, trade shows, and other web sites. This final chapter will introduce you to some of these valuable resources including more tips on AutoCAD.

Profiles

You may find that you would like to have different toolbars for different types of work. Or, if you work in a multiple-user computer lab such as in an educational setting, there might be a need to keep systems settings (such as toolbars) to a standard format and then allow users to select individual settings. The Preferences dialog box allows you to do this with the Profiles tab (Fig. 17-1).

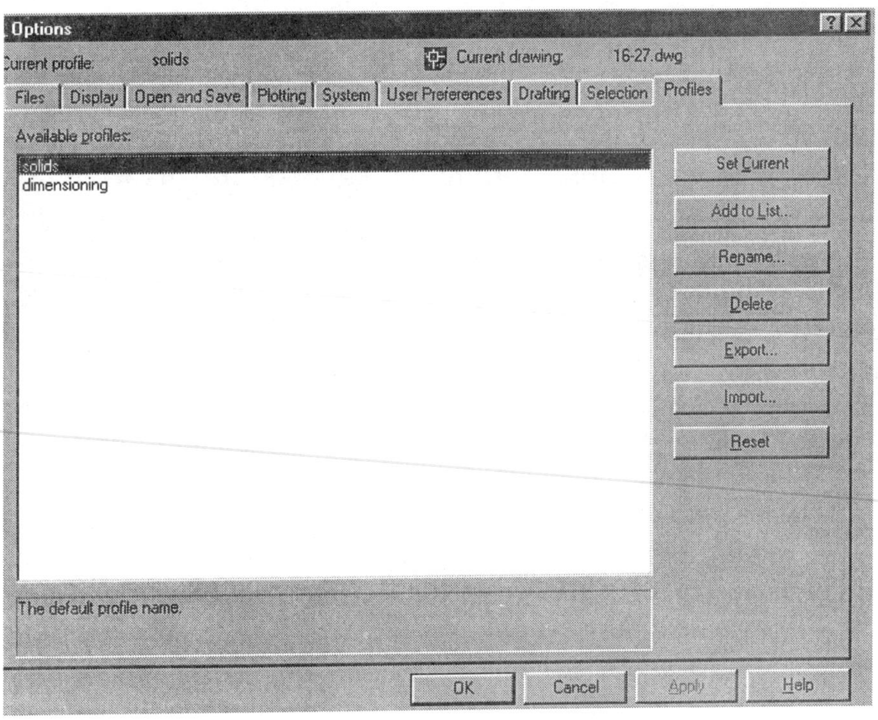

Fig. 17-1 Profiles

1. Menu Bar: Tools > Options.
2. Click on "Profiles" tab.
3. To create your own, copy the selected profile to a new name.
4. Modify the screen (add or delete toolbars).

Your changes are automatically saved to the new name. To change the profile, enter:

1. Menu Bar: Tools > Options.
2. Click on "Profiles" tab.
3. Select the other profile listed.
4. Click on "Set Current" button.

Using AutoCAD Drawings in Other Programs

There are several methods for saving images of AutoCAD drawings to be placed in in documents created by other software such as Adobe Pagemaker or Microsoft Word. One method is is to export the drawing to one of many different formats such as a bitmap or DXF. You will need to know what format the other program requires. The procedure is listed below.

1. Menu Bar: File > Export...
2. Change the file format to "DXF."
3. Enter a file name (which can be the same as the drawing name).

Another method is to save a graphic image of the drawing as it appears in the active viewport. The procedure is listed below.

1. Menu Bar: Tools > Display Image > Save...
2. Click on "OK" to save in a BMP format.
3. Type name for image file and click Save to close window.

You can also save to a "TIF" or "TGA" format with additional options.

Still another method is to use a screen capture utility. This has the advantage of showing the screen menus and toolbars. These are available on the internet at a reasonable price.

Placing Images in AutoCAD

In the same way, you can insert images or drawings from other programs into your AutoCAD drawing. The typical image will be a bitmap or the "BMP" format:

> **1. Menu Bar: Insert > Raster Image...**
> **2. Browse and find a "BMP" file to insert.**

The raster image can be scaled, rotated, or moved but other edit commands cannot be used on a raster image.

HyperLinks in Drawings

You can insert hyperlinks by using the icon. You must specify web site address or URL or alternatively you could specify a document on your hard drive. This might be used where you place a manufactuer web site hyperlink on a furniture piece in plan. When someone clicks on the hyperlink, an internet browser is launched and the company's web site perhaps on the specific page with information on the furniture piece is given. Another idea might be that you insert a hyperlink on a floor plan that launches a walk-through animation of the interior that was created with 3D Studio Viz.

DWF Files

You can post drawings on the internet with the DWF format. Then, if your reader has the WHIP plug in, they will be able to zoom and pan the drawing. DWF files are created when you plot to a DWF format. These procedures are beyond the scope of this beginning tutorial.

Autodesk Web Site

Autodesk, developer of AutoCAD and other CAD software, has an extensive web site with useful information on products, AutoCAD training centers, and publications. The site also offers discussion groups related to the products so you can have your questions answered by another CAD user or staff at Autodesk. You will want to bookmark this site for frequent updates:

> **www.autodesk.com**

AutoDesk Architectural Desktop

Many add-on software packages are available which can customize AutoCAD for a variety of disciplines and fields. Architectural Desktop by Autodesk is specifically oriented toward architecture and interior design. With its object-oriented technology, you can automaticaly insert doors and windows and generate walls based on a space layout plan. The software adds construction documentation including schedules and reports.

Manufacturer Symbol Libraries

Furniture manufacturers offer symbol libraries to use with AutoCAD. These are usually free to designers. A few, such as Herman Miller, are available directly from their web site while others must be obtained from your local furniture dealership. Most of the symbol libraries replace the standard AutoCAD menu with an enhanced menu that allows you to access the symbols either through the menu bar or a right screen menu. Some of these manufacturers and their web site addresses are listed below.

Haworth	**http://www.haworth.com/**
Herman Miller	**http://www.hermanmiller.com**
Steelcase	**http://www.steelcase.com/**

A/E/C Systems

A/E/C Systems, an annual trade show and conference, showcases the latest in design tools and equipment for the architecture, engineering, and construction industry. Autodesk Expo, an extensive part of the show, has many product demos and educational sessions. The show is held during the month of June, varying its location in major cities in the United States. Admission to the trade show is free if you register early. Information on this valuable show is available from the web site:

www.aecsystems.com/

Books/CD-ROMS

Finkelstein, Ellen. *AutoCAD 2000 Bible.* IDG Books, 2000.

This book contains a variety of drawing examples and advice. The CD-ROM contains drawings, tutorials, symbol libraries, and shareware.

Leach, Jim *AutoCAD 2000 Instructor.* McGraw-Hill, 2000.

These tutorials are well organized and detailed, providng an excellent reference and tutorials for the designer. Chapters are grouped around related commands so that specific topics can be studied. Teachers will appreciate this book as an additional reference.

Omura, George. *Mastering AutoCAD. 2000* Sybex, 1999.

This best-selling book is comprehensive and covers advanced topics, such as VBA programming, Internet distribution, and 3D modeling. Many design firms use this book as their primary desk reference on AutoCAD.

Smith, Bud, Middlebrook, Mark, & Smith, Bud. *AutoCAD 2000 for Dummies: A Reference for the Rest of Us.* IDG Books, 1999.

Some designers like the casual tone of this book while others might find it annoying. If you like the other Dummy books, you will probably like this one too.

Stellman, T.A. Krishnan, G.V. *Harnessing AutoCAD 2000.* Autodesk Press/ Thomson Learning, 2000.

This comprehensive reference is a step-by-step tutorial to learning AutoCAD. Starting with basic commands, it builds on competencies similar to this tutorial. Detailed explanations of all the basic commands make it a useful desktop reference.

Index

E

F

G

H

I

L

M